FLOWERING TREES IN SUBTROPICAL GARDENS

FLOWERING TREES IN SUBTROPICAL GARDENS

GÜNTHER KUNKEL
Drawings by Mary Anne Kunkel

Dr. W. Junk b.v., Publishers, The Hague–Boston–London 1978

ISBN 90 6193 592 x

Cover design Max Velthuijs

CONTENTS

AN APOLOGY

Any study on flowering plants for or in gardens is a never-ending pleasure as new 'treasures' are continually being introduced from elsewhere or are there to be discovered in some forgotten corner. It is an uncomplicated pleasure as most species are grown for their beautiful flowers, their remarkable foliage, their prized fruits or other outstanding properties; it is also a rather safe pleasure as 99.9% of these plants are known and have been duly described in every botanical sense. We are able to keep to firm ground far from the field of priority matters and other causes of stress, ungrudgingly accepting that a flower is a flower is a flower ...

Trees make no exception. Although scientists may growl over a correct name or over the acceptance of a plant as a Tree because its trunk is not woody enough to be called 'wood', we may forget about all this and join the numerous groups of flower-admirers and 'beauty-watchers', even if the flowers of our most beloved are rather inconspicuous. Their real value can never fit tidily into any established order but depends on the individual relationship between an object and its admirer. Because rôles are created by the logical mind of man whereas praise and admiration belong to the realm of the senses.

There are over 20.000 species of flowering plants which, according to their height, habit and other basic characters should be considered as 'trees'. What *is* a tree? – According to the Royal Horticultural Society's *Dictionary of Gardening* and other books, a tree is supposed to be 'a woody perennial rising from the ground with a distinct trunk'. One may certainly agree with this definition in principle; however it is too simplified as it leaves out palms and dragon trees, large cacti and euphorbias, and even the venerable Ombú of the Argentinian pampa is not woody but classed as 'a giant herb'. On the other hand some *Ononis* and *Plantago* species, for example, do have a short woody trunk, but are they trees? *Erica arborea*, the Tree Heath, is a proper tree while growing in its native forest community but remains a shrub after having been cut or where growing outside its optimal or climax formation; and the Oleander, by pruning, may be converted into a tree. Our definition remains relative.

The German dendrologist Professor Schmucker while asking 'What is a

Tree?' tried to resolve the question after a 5-page discussion by stating that 'a tree is a high and upright terrestrial plant'. – Edwin A. Menninger ('The Flowering Tree Man'), A. Mitchell and others have asked the same question, and we may repeat it over and over again. Supposing that everyone knows what is a 'bush', I suggest that *a tree is a long-living woody or fibrous plant, with a single or multiple trunk, in its mature stage and in a suitable environment larger than a bush.*

Most trees provide wood: firewood and charcoal, wood for constructions and making both large and small implements, for furniture, houses, bridges and shipbuilding. Trees are mythical, are sacred to certain 'primitive' tribes or may even cause inexplicable and archaic feelings of fear. Some people are buried in coffins made of wood from trees, others plant a tree after a child has been born to them. Trees give shade and shelter, they attract clouds and moisture, and many bear edible fruits. Trees belong to our natural surroundings. Landscapes deprived of trees are wastelands: an un-desirable inheritance.

But trees are also beings, with their own feelings and their language, willing to give when well-treated, silently suffering under adverse circumstances and, so they say, able to cry when brutally cut down. We had better accept them as part of our family.

The very serious reviewer is asked to forgive this rather unorthodox introduction, but be assured, trees know where they belong. Of course sometimes they make mistakes. We usually do.

INTRODUCTION

In 1969 volume I of 'Arboles Exóticos' appeared in Las Palmas de Gran Canaria, published in Spanish by the Island Council (Cabildo Insular). Volume II was to have dealt with a further 100 or so Dicotyledons, and volume III to have closed the cycle with Gymnosperms and the tree-like Monocotyledons. Neither of the latter having been published it was suggested that a totally revised and enlarged version of the first book be prepared for Dr. W. Junk, Publishers, The Hague. We are most obliged to Mr. S. P. Bakker and the Board of Directors for their interest in the present work, hoping that it will fill a gap and help both residents and visitors to get to know the amazingly rich exotic flora that is to be encountered in Mediterranean and Canarian parks and gardens. A second volume is in preparation and will concentrate on the bushes and shrubby trees left out in this one.

In *Flowering Trees in Subtropical Gardens* special attention is given to species found in Canary gardens. As, however, most trees described are widely distributed in other regions with a similar Mediterranean climate, it is hoped that this guide may be of use in gardens of the subtropics in general. Several of the species selected here are little known in gardens, hardly ever found illustrated in current reference books and have therefore been included for interest's sake.

This book is written for 'Everyman', be he (or she) gardener, scientist or so-called layman, indeed for all who are interested in cultivated trees. Scientific terms and detailed descriptions have been avoided whenever possible; but for further information and for cross-checking purposes ample references are given for each species under consideration. Where an author's name appears in brackets, e.g. (Uphof), it is because he mentions or describes a certain species under a different name, a name then considered here as a synonym.

Most species are illustrated, either in full or in part, by line drawings which accompany and complement the text. The drawings have been made from living material where ever possible, or from dried herbarium specimens; all work was compared with available literature.

Initial recognition of a particular tree may be achieved by first consulting the classified tree list and the simplified key (based mainly on leaf characteristics). Latin names are used in both cases but an index to the common names is given in a chapter following the descriptive part.

No excuse is offered for having presented some rather voluminous genera such as *Acacia*, *Ficus* and *Eucalyptus*; they are so numerous in gardens, in parks and along roadsides that we feel they really deserve the attention given them. On the other hand certain world famous flowering trees such as

Adansonia digitata, *Anacardium occidentale*, *Cordia sebestena*, *Dalbergia sissoo*, *Dillenia indica*, most Proteaceae, *Tectona grandis* and others, have been left out owing to lack of local material.

Taking the Canary Islands as an example: It is a region that for several centuries has been an Atlantic cross-road for travellers, who have often brought with them the rare and strange plants encountered on their far-flung journeys. The benign climate of these islands has allowed most of these plant tourists to settle down and become resident. The resulting cosmopolitan plant population is best observed in the public gardens of the larger towns and cities. A walk through the so-called garden cities can also be extremely rewarding, for many gardens harbour a wealth of species which is quite astonishing. Most town gardens are small and many plants obligingly overflow into the streets. But most of all, if access can be obtained to some of the older private estates, here the richness and diversity of plant collections are on as high a level as any smaller botanical garden. The roadside flora in the hinterland is also of much interest.

As pointed out in the main text some exotics have become naturalized, which is to say, they seed and spread without human aid. Other plants being either poisonous and/or prickly should be treated with the respect due to them.

In this book a total of over 150 species, from 98 genera, belonging to 48 families are dealt with. The Palms and Conifers have had to be excluded for there are so many of them that they would have made the book too voluminous; so we have kept strictly to the flowering exotic trees. Many native trees are mentioned in the introductory notes on families, and titles of works concerned with native species are given in the Bibliography.

Acknowledgements

We wish to thank The Director and Staff of The Royal Botanic Gardens, Kew, for all their help in sorting out nomenclatural problems. Our special thanks to Mr. Alfred Hansen (Copenhagen) and Dr. Hervé M. Burdet (Geneva), for providing photocopies and useful information on certain species. We are also grateful to Prof. E. J. H. Corner and Dr. Gorden P. De Wolf (*Ficus*), and Prof. G. Moggi (*Eucalyptus*), for their assistance in naming material. And to our many gardener friends for allowing us to invade their gardens, to study specimens and to keep material of interest. Last but not least we must not forget to thank fellow authors for their valuable publications from which we have taken the liberty to quote with evident freedom.

The authors wish to dedicate this little 'dendro-guide' to their friends Alfred Hansen (Copenhagen, Denmark), Joachim Illies (Schlitz, Germany) and David McClintock (St. Mary Platt, England) thanking them herewith for their support and interest in this project. Also to our friends in the Canary Islands who try against all odds to keep trees as their companions.

G. & M. A. Kunkel

GLOSSARY

acorn: the fruit of an Oak, an ellipsoid nut partly covered by the enlarged involucre or cupule.
acuminate: from acumen (pointed); tapering into a narrow point.
apex: the tip (of a leaf).
aril: from arillus, dry or fleshy cover or bedding of seeds, often brightly coloured.
armed: from armatus = provided with thorns, spines or other kinds of 'prickles'.
berry: a fleshy fruit usually containing several seeds.
bipinnate: twice pinnate, a compound leaf of which primary divisions are also pinnate.
blade: or lamina, the expanded or laminar part of a leaf.
bract: leafy structure at the base of a flower or flowerhead, often coloured and can be mistaken for a 'flower'.
buttresses: plank-like supports at the base of trunks.
calyx: the outer envelope of a flower composed of the sepals.
campanulate: bell-shaped.
capsule: a dry fruit formed by two or more segments usually splitting open at maturity.
catkin: an amentum; unisexual flower-spike (as in *Salix*).
cauliflor: cauliflorous, bearing flowers on the trunk and/or main branches.
constricted: contracted, or compressed in between seeds as in the pods of some *Acacia*.
cordate: heart-shaped.
coriaceous: leathery, thick, tough.
corolla: group (free or united) of petals which forms the flower, usually of distinctive colour.
corymb: flowerhead (inflorescence) with stalked flowers originating at different levels but forming a flattened head.
crenate: with rounded teeth, or scallops in a leaf-margin.
cuneate: wedge-shaped, tapering towards the base.
cupule: the cup-shaped involucre partly covering a nut (acorn).
cv.: cultivar, a variety originated in cultivation.

cyme: a branched inflorescence of which the central flower opens first and stops further development of this particular part.
deciduous: or caducous; falling off, not persistant during a certain time of the year.
decussate: alternately crossed in opposite pairs.
dehiscent: when referring to a fruit = opening at maturity; capsules are usually dehiscent.
deltate: deltoid, forming a triangle.
dentate: toothed, with sharp teeth on the edge (margin).
diameter: ∅; in the present work referring to the trunk (stem) of a tree at breast height.
digitate: a compound leaf with usually 5 or more leaflets diverging from a common point (like the fingers of a hand).
dioecious: 'two houses', having male and female flowers on separated individuals.
downy: pubescent, covered with fine short hairs.
drupe: a 'stone-fruit' usually having a single seed covered by the fleshy pulp (i.e. Cherry, Peach, Plum etc.).
ellipsoid: rounded but not circular; a body with an oval outline.
elliptic(al): oval in outline or twice as long as wide.
emarginate: shallowly notched at the apex; deeply notched or cut leaves as in *Bauhinia* might not be truly emarginate but two leaves joined half-way or further up.
entire: an even leaf-margin, unbroken by teeth, lobes etc.
evergreen: always green, not deciduous; leaf change taking place gradually, little noticed.
falcate: curved, sickle-shaped.
fig: a fleshy, more or less globular fruit composed of a hollow receptacle with monoecious flowers on the inner surface.
follicle: usually two pod-like capsules on a common stalk, set at an angle to each other (i.e. in *Plumeria*).
funicle: cord-like filament partly surrounding a seed and being helpful in its distribution.
glabrous: glabratus, smooth, not hairy.
glands: small bodies secreting oils, acids, latex etc., single or few-celled, usually on leaf-stalks.
glaucous: grey-waxy or bluish-green.
globose: spherical, round, ball-shaped.
heterogenous: not uniform; of several sizes, shapes etc.
heterophyllous: having leaves of different shapes.
imparipinnate: odd or unevenly pinnate, having a terminal segment or leaflet.
indehiscent: when referring to a fruit = not opening when ripe.
inflorescence: the arrangement of the flowers on a common stalk.
lacerate: laceratus; torn, rather unequally cut.
lanceolate: lance or spear-shaped, narrowly elliptic and tapering at both ends.
latex: milky fluid in glands or veins (as in *Ficus*); not necessarily only white.

leaflet: segment or part of a compound leaf.
lenticels: corky spots on the bark which act to aid aeration of underlying tissue.
linear: linearis; very narrow, with parallel edges.
marcottage-system: (or 'mossing'), system employed to obtain rooted cuttings, especially from fig trees; a part of a branch is wrapped in moist peat or moss and covered with plastic. The branch section can be cut off (below the mossing) once rooting has taken place.
margin: edge of a leaf.
midrib: the main or central vein of a leaf.
monoecious: flowers or inflorescences of separate sexes but occurring on the same plant.
mucronate: abruptly ending in a sharp point.
nerves: veins (of a leaf).
nerve system: nervation, venation; arrangement of nerves.
noxious: harmful, injurious.
nut: indehiscent, usually hard-shelled fruit with one single seed.
oblanceolate: inverted or inversely lanceolate.
obovate: inversely ovate with the broadest part uppermost.
operculum: a lid; operculate = covered by a lid.
orbiculate: circular, disk-shaped.
palmate: from palma = hand; palmate leaves = leaf-segments radiating from a centre point near to the base.
palmatisect: in the case of leaves they are cut almost to the base.
panduriform: fiddle-shaped.
panicle: a branched inflorescence, each branch bearing several flowers.
paripinnate: evenly pinnate, without a single terminal leaflet.
petals: separate or united, fine-leafy, usually colourful segments forming the corolla.
petiole, petiolate: leaf-stalk, stalked.
phyllodes: leaf-like petioles or stalks, flattened and narrow, performing the functions of leaves (i.e. in several Acacias).
pinnae: (sing. pinna), name given to the primary divisions of a bipinnate leaf; segments of a pinnate leaf.
pod: usually a dry fruit with several seeds; also known as legume or siliqua.
polymorphic: of several or many forms.
pubescent: from pubens = downy.
raceme: inflorescence of stalked flowers on an unbranched axis.
rachis: from rhachis = axis, of a flowerhead or a compound leaf.
rhomboid: rhombic, diamond-shaped; a square or oblong surface tipped at a lopsided angle.
samara: a winged, indehiscent fruit (i.e. of *Acer*).
scaly: with scales, small leafy parts or appendages.
serrate: margin with small sharp teeth = saw-edged.
sessile: sitting, not stalked.
sinuate: with a strongly recessed margin.

spathulate: spatula-shaped; broad at the apex and soon tapering towards the base.
spike: unbranched inflorescence with sessile flowers.
spine: not only a 'prickly' but also a difficult problem to resolve as Latin and Germanic languages use opposite definitions. A spine makes a thing spinous, but something prickly may have all or anything that 'pricks'.[1] In short and as employed in the present work: a sharp-pointed, stiff, usually straight body which may be a modified stipule, leaf or branchlet, and which is not detachable without damage to the cellular structure.
s.str.: sensu stricto, in the strict sense.
stamen: male reproductive flower-organ bearing pollen sacs protruding from or protected by the corolla.
stilt roots: stilted or leggy roots especially common in plants growing in very wet or swampy places.
stipule: scale-like appendages, typical in leaf-axils of certain plant groups; often deciduous, occasionally converted into spines.
sub: nearly, under, slightly; subcoriaceous = almost leathery; subspecies = nearly (or just below) a species.
syncarp: from syn = united, joined; in syncarps a multiple compound fruit (i.e. *Morus*).
taxon: scientific unit of classification (i.e. variety, species, genus, family); taxonomists = scientists occupied with such matters.
thorn: a prickly superficial, often hooked and not woody appendage, easily detachable. For further discussion see 'spine'.
tomentose: having a tomentum = covered by dense, short rigid hairs.
trifoliate: compound but with only 3 leaflets.
truncate: very abruptly ending as if cut off.
undulate: wavy; usually applied to a leaf-margin alternately concave and convex.
unequal: disequal, uneven; not of equal size and shape.
unisexual: of one (or the other) sex, not together in the same flower.
urceolate: pitcher- or urn-shaped.
variety: subtaxon of a species.
veins: fibro-vascular system, main conductors of plant sap.
velvety: from velvetum = densely covered by short hairs.
verticillate: (leaves or branches) in whorls.
winged: winged fruit, winged rachis = provided with or partly covered by a laminar segment.

[1] If the German *Dorn* is rightly translated as *spine*, then a Rose is spiny, a fact which would call for the revision of numerous folk-songs and other lyrical exclamations. The late but still much consulted Spanish botanist P. Font Quer ('Diccionario de Botánica', Editorial Labor 1953) states that the English '*spine* corresponde a *aguijón* (thorn), y la *espina* (spine) se llama *thorn*'. – A rather confusing situation often overcome by using the epithet *Prickle:* 'a sharp-pointed, irregularly placed outgrowth of the surface of a plant-part' (Barrett 56). Supposing that – until further agreement – the reader accepts occasional spines and thorns as 'prickles', he is to be reminded that some prickles might be more prickly than others.

CLASSIFIED TREE LIST

Trees with conspicuous flowers

a. outstanding

Brachychiton spp.
Cassia spectabilis
Catalpa bignonioides
Chorisia speciosa
Delonix regia
Dombeya spp.
Erythrina spp.
Grevillea spp.
Hibiscus elatus
Jacaranda mimosifolia
Koelreuteria paniculata
Lagerstroemia speciosa
Lagunaria patersonii
Magnolia spp.
Melia azedarach
Metrosideros excelsa
Pachira insignis
Plumeria rubra
Spathodea campanulata
Tabebuia rosea
Tipuana tipu

b. notable

Acacia spp.
Amygdalus spp.
Arbutus unedo
Azadirachta indica
Bauhinia forficata
Citharexylum spinosum
Citrus spp.
Duranta repens
Erica arborea
Eucalyptus spp.
Guaiacum officinale
Laurocerasus lusitanica
Macadamia integrifolia
Mammea americana
Parkinsonia aculeata
Pittosporum spp.
Robinia pseudacacia
Stenocarpus sinuatus
Thespesia populnea
Thevetia peruviana

Fruit and nut trees

a. exclusively utilitarian

Amygdalus spp.
Annona spp.
Armeniaca vulgaris
Carya illinoensis
Casimiroa edulis

b. dual-purpose

Aleurites moluccana!
Arbutus unedo
Ceratonia siliqua
Coccoloba uvifera
Corynocarpus laevigata!

Castanea sativa
Citrus spp.
Diospyros kaki
Eriobotrya japonica
Ficus carica
Juglans regia
Macadamia integrifolia
Mangifera indica
Olea europaea
Persea americana
Psidium spp.
Dovyalis hebecarpa
Ficus spp.
Mammea americana
Melicoccus bijugatus
Morus spp.
Pouteria campechiana
Syzygium spp.
Tamarindus indica
Terminalia catappa

! = partly poisonous

Species with other uses

Cinnamomum camphora (medicinal)
Cinnamomum zeylanicum (spice)
Eucalyptus globulus (medicinal)
Quercus suber (cork)
Sapindus saponaria (soap)
Tilia platyphyllos (medicinal)

Poisonous (or partly poisonous) trees

Aleurites moluccana
Corynocarpus laevigata
Duranta repens
Erythrina spp.
Hura crepitans
Ilex aquifolium
Laurocerasus lusitanica
Melia azedarach
Plumeria rubra
Sapindus saponaria
Thevetia peruviana

Invading or aggressive species

Acacia spp.
Ailanthus altissima
Albizia lophantha
Celtis australis
Eucalyptus spp.
Gleditsia triacanthos
Populus spp.
Robinia pseudacacia
Ulmus minor

KEY (mainly based on leaf-characters)

Leaves reduced, or modified

Leaves reduced to minute 'scales' found at the nodes of needle-like flexible twigs; flowers inconspicuous **Casuarina** spp.

Leaves needle-like, stiff

leaves very short, more or less whorled; flowers small, whitish, bell-shaped **Erica arborea**

leaves very long, flowers showy, greenish-yellow, in upright terminal spikes **Grevillea nematophylla**

Leaves modified to phyllodes, spineless; flowerheads spherical, yellow to golden yellow; fruits narrow pods see **Acacia** spp.

Leaves simple, latex present

Latex white, leaves alternate

leaves persistant, usually entire

of various shapes and colours; flowers hidden in a hollow receptacle becoming a fig **Ficus** spp.

very narrow; flowers yellow, rather showy; fruit a kind of a swollen bag **Thevetia peruviana**

leaves deciduous, entire or lobed

branches spiny, leaves entire

fruit a spherical syncarp **Maclura pomifera**

branches unarmed

leaves toothed or lobed

flowers open; fruit agglomerated (syncarp) **Morus** spp.

flowers hidden in a hollow receptacle **Ficus carica**

leaves entire, oblong-lanceolate

flowers showy; fruit a follicle **Plumeria rubra**

Latex yellowish; leaves opposite or alternate

leaves opposite, large, fine-veined; flowers showy, white; fruit a globose berry **Mammea americana**

leaves alternate, narrow-oblong, strong-veined; flowers greenish, little conspicuous; fruit egg-shaped **Pouteria campechiana**

Latex viscid; leaves alternate

leaves lobed, maple-shaped; flowers white; fruits in bunches, nut-like **Aleurites moluccana**

leaves ovate to heart-shaped; flowers purplish, very small; fruit a strong-ribbed subglobose capsule **Hura crepitans**

Leaves simple, latex absent

Leaves alternate

branches usually spiny or with spiny stipules

leaves entire, sometimes with a winged rachis; flowers large, white; fruit globose, juicy, edible **Citrus** spp.

branches not spiny but leaves usually with a spiny margin

flowers small, white; fruit spherical, less than 1 cm ∅, scarlet, not edible **Ilex aquifolium**

flowers greenish, inconspicuous, clustered or in narrow catkins; fruit an acorn

bark of trunk corky **Quercus suber**

bark of trunk not corky **Quercus ilex**

branches and leaves unarmed

leaves very narrow (see also Erica, Casuarina, Grevillea nematophylla, Acacia, Thevetia)

leaves usually deciduous, up to 1,5 cm wide; flowers in cylindrical catkins **Salix alba**

leaves persistant, up to 0,5 cm wide; flowers solitary, cream-coloured **Pittosporum phillyraeoides**

leaves elliptical, ovate, obovate or lanceolate (see also Ficus, Mammea, Pouteria, Citrus)

margin entire, usually regular

leaves oblong-elliptic, pale tomentose beneath; flowers pink, showy; fruit a globose capsule **Lagunaria patersonii**

leaves oblong-obovate, rusty downy beneath; flowers large, white or pink; fruit cone-shaped **Magnolia** spp.

leaves oblong to lanceolate, more or less glabrous on both sides flowers rather inconspicuous, greenish or white;

fruits small, inconspicuous **Cocculus laurifolius**

fruit a fragrant, orange-coloured drupe **Corynocarpus laevigata**

fruit bluish-black, olive shaped **Cinnamomum zeylanicum**

fruit large, a fleshy or fibrous drupe **Mangifera indica**

fruit very large, a kidney-shaped fleshy, berry-like aggregate **Annona** spp.

flowers conspicuous, solitary or grouped, white or cream-coloured

flowers labiate-like, cream-coloured; fruit greenish, olive-shaped **Bontia daphnoides**

flowers regular, white and fragrant; fruit globose, juicy, orange or greenish **Citrus** spp.

flowers whitish (or red), many-stamened; fruit a bowl-shaped capsule **Eucalyptus** spp.

leaves oblong-elliptical, glabrous, with or without glands; flowers greenish

flowers solitary or in clusters; fruit a large, fleshy berry-like aggregate **Annona cherimolia**

flowers small, in larger panicles

leaves fragrant, glands present; fruit almost black, olive-shaped **Cinnamomum camphora**

leaves not fragrant, glands absent; fruit a large fleshy drupe **Persea americana**

flowers conspicuous, whitish, in showy racemes

flowers with protruding stamens; fruits small fleshy ribbed berries **Phytolacca dioica**

flowers bell-shaped; fruit a woody, dark brown boat-shaped capsule **Brachychiton diversifolium**

leaves obovate or oblanceolate, clustered

leaves very large, almost whorled; flowers in narrow spikes; fruit an almond-like nut **Terminalia catappa**

leaves oblong-obovate; flowers whitish, in clusters; fruit a capsule up to 1,5 cm long **Pittosporum tobira**

leaves oblanceolate to subspathulate; cauliflorous; fruit a large 'calabash' **Crescentia cujete**

margin undulate
leaves evergreen, about lanceolate; flowers white or yellowish
flowers in clusters; fruit a small capsule **Pittosporum undulatum**
flowers in racemes; fruit a nut **Macadamia integrifolia**
leaves deciduous or semi-persistant, oval-shaped to elliptic; flowers greenish
fruit a purple velvety berry, 2,5 cm ∅ **Dovyalis hebecarpa**
fruit orange-coloured or reddish, smooth-skinned, tomato-shaped, 5–7 cm ∅ **Diospyros kaki**

margin dentate or serrate
leaves ovate or oblong, pointed, base unequal; flowers greenish, inconspicuous

fruit a black, thin-fleshy drupe **Celtis australis**
fruit winged, papery, yellow **Ulmus minor**
leaves roughly lanceolate, base cuneate; flowers conspicuous
flowers in spikes or racemes
flowers (male!) pale green, in narrow spikes; fruit globose, prickly, with brown shiny chestnuts **Castanea sativa**

flowers white, in showy racemes; fruit small, ovoid, fleshy, almost black **Laurocerasus lusitanica**

flowers in showy panicles, white or pink
leaves up to 25 cm long, very stiff; corolla open, white; fruit a fleshy, yellowish, soft-skinned berry **Eriobotrya japonica**
leaves up to 8 cm long, leathery but not stiff; corolla pitcher-shaped, pinkish; fruit a scarlet, rough-skinned mealy berry **Arbutus unedo**

flowers solitary or in pairs, large, white to pinkish; fruits velvety

leaves lanceolate, subcoriaceous
fruit 2–2,5 cm ∅, little fleshy, when drying leaving only a nut **Amygdalus communis**

fruit fleshy and juicy, rose-coloured, up to 9 cm ∅ **Amygdalus persica**
leaves roughly elliptical, herbaceous
fruit fleshy and juicy, yellowish, up to 5 cm ∅ **Armeniaca vulgaris**
leaves maple- or heart-shaped or of other forms (see also Hura)
leaves large, roughly maple-shaped, lobed; fruits usually 2 spherical heads 3–4 cm ∅ **Platanus × hybrida**

leaves smaller, triangular to rhomboid, less lobed; fruits small but showy catkins

leaves somewhat lobed, white-tomentose beneath **Populus alba**
leaves dentate-sinuate, grey-tomentose beneath **P. canescens**
leaves serrate, glabrous beneath **P. nigra**

leaves roughly heart-shaped, entire

short-stalked, wider than long, leathery; flowers inconspicuous in narrow racemes; fruit drop-shaped, a purple fleshy berry **Coccoloba uvifera**

long-stalked, herbaceous to half-leathery; flowers very showy but short-lived; capsule 2–3 cm ∅

leaves as long as wide, short-pointed; petals narrow, curled, orange turning purple; capsule tomentose **Hibiscus elatus**
leaves longer than wide, long-pointed; petals broad, not curled, pale yellow turning purple; capsule smooth **Thespesia populnea**

leaves roughly heart-shaped, serrate-dentate

leaves large, somewhat unequal, up to 10 cm long and wide; flowers whitish; fruits small, tomentose, hanging on stalks with a partly attached laminar bract **Tilia platyphyllos**
leaves equal, up to 20 cm wide, soft-hairy; flowers pink in showy long-stalked 'balls' **Dombeya × cayeuxii**

leaves more lobed, smaller; flowers white, solitary or grouped **Dombeya tiliacea**

leaves sinuate or obtusely lobed; flowers small, greenish; fruit an acorn **Quercus robur**

leaves irregular: entire, 3 or 5-lobed to almost pinnatipartite

flowers bell-shaped, red, in large showy panicles; fruit long-stalked, woody, boat-shaped **Brachychiton acerifolium**

flowers comb-shaped, scarlet; fruit a small flattened capsule **Stenocarpus sinuatus**

Leaves opposite or decussate

leaves opposite, palmately lobed; fruits winged, in pairs (samarae) **Acer pseudoplatanus**

leaves roughly in whorls, broad-ovate to heart-shaped, very long stalked; flowers about bell-shaped; capsule narrow, black, up to 35 cm long **Catalpa bignonioides**

leaves decussate but usually in a flattened position

blade entire

leaves narrow, stiff; flowers white, small; fruit the proper olive **Olea europaea**

leaves larger, usually persistant

blade herbaceous, ovate-elliptic, 5 cm long; flowers white or purple, in terminal racemes; fruit small, fleshy, orange-coloured **Duranta repens**

blade herbaceous while immature, of different shapes, later alternate and coriaceous **Eucalyptus** spp.

blade subcoriaceous or coriaceous

oblong, pubescent beneath; stamens crimson, in showy heads; capsule almost bell-shaped **Metrosideros excelsa**

obovate, glabrous; stamens white; fruit a fleshy berry 2–2,5 cm ∅ **Psidium cattleianum**

oblong-ovate, pointed; flowers small, white, in showy panicles; fruit small, fleshy, bluish-black **Ligustrum lucidum**

lanceolate; with edible fruits

flowers large, clustered, long cream-coloured stamens; fruit a brownish globose berry up to 4 cm ∅ **Syzygium jambos**

flowers small, white, in branched inflorescences; fruit lilac-purple or black, roughly plum-shaped **Syzygium cuminii**

leaves large, deciduous

blade roughly oval-shaped, often folded

flowers axillary, white; fruit globose, yellow-skinned, fleshy, up to 7 cm ∅ **Psidium guajava**

flowers in showy terminal panicles, pink; fruit capsule woody, almost spherical, up to 2,5 cm ∅ **Lagerstroemia speciosa**

leaves larger, branches angled; flowers small, white, very fragrant, in long terminal racemes; fruits small, fleshy, globose, orange-coloured or purple **Citharexylum spinosum**

blade with a serrate margin

leaves lanceolate; flowers small, greenish; fruit green, fleshy, broad olive-shaped **Elaeodendron orientale**

Leaves pinnate, digitate or trifoliate

Leaves digitate, leaflets 3 to 8

leaflets 5 to 8; trunk strongly armed

flowers large, rose-coloured; capsule oblong or pear-shaped, up to 7 cm ∅ **Chorisia speciosa**

leaflets 4 to 7; trunk not armed
flowers with crimson petals and bunches of protruding stamens; capsule globose, 15–20 cm ∅ **Pachira insignis**

leaflets 3 to 5, rarely single
flowers whitish-pink, trumpet-shaped; capsule linear-cylindric, up to 20 cm long **Tabebuia rosea**

flowers small, greenish-white; fruit globose, fleshy, up to 7 cm ∅ **Casimiroa edulis**

Leaves imparipinnate
leaflets entire, 3–5 (usually 3), glands present; fruit a pod
bark corky, terminal branches usually drying off; flowers crimson-scarlet, in racemes 30–50 cm long **Erythrina crista-galli**

bark not corky, branches not drying off
branches strongly armed; flowers broad, red to scarlet, in short clusters or racemes **Erythrina caffra**

branches inconspicuously armed; flowers narrow, scarlet, in narrow racemes 20–30 cm long **Erythrina variegata**

leaflets coarsely dentate, 3–5 (rarely 7), glands absent
flowers yellowish-green, in pendent racemes; fruits winged (samarae) **Negundo aceroides**

leaflets many, usually opposite
fruit a nut; flowers very small, greenish
leaflets many, lanceolate, unequal, serrate; nut ovoid, splitting **Carya illinoensis**

leaflets 5–9, ovate, regular, entire; nut oval-shaped, remaining closed **Juglans regia**

fruits winged; flowers showy or not
fruits in opposite pairs
leaflets 3–5 (rarely 7) **Negundo aceroides**
fruits single, or on a common stalk
leaflets 5–9; flowers inconspicuous **Fraxinus angustifolia**
leaflets many; flowers in large bunches
leaflets oblong, entire; flowers yellow **Tipuana tipu**
leaflets oblong-lanceolate, irregularly toothed, often with glands; flowers whitish-green **Ailanthus altissima**

fruits capsules or pods; leaflets ovate to oblong-lanceolate
leaflets entire, oblong or oval-shaped
flowers white, usually in pendent racemes; pods flat, brown **Robinia pseudacacia**

flowers orange-red, in upright crown-like clusters; capsule flattened, splitting in 2 valves **Spathodea campanulata**

leaflets serrate or lobed, ovate-lanceolate
flowers yellow, in large panicles; capsule short, inflated, 3-valved **Koelreuteria paniculata**

fruits of other shapes; flowers paniculate
leaflets ovate or obovate, entire; flowers inconspicuous; fruits very small scarlet drupes **Schinus terebinthifolius**

leaflets of variable shape, entire or somewhat spiny; flowers irregular cup-shaped, dark purple; fruit cylindrical, sausage-like **Kigelia africana**

leaflets oblong-lanceolate, coarsely dentate; flowers white; fruits ovoid berries, yellow or brown **Azadirachta indica**

Borderline cases, to be investigated
leaves large, 'fern-like', narrowly lobed or dissected; flowers rich golden, in comb-like racemes; capsules small, dark, flattened **Grevillea robusta**

leaves deeply emarginate; probably 2 joined leaves, with numerous stipules (reduced leaflets?); flowers with long, narrow white petals; pod woody, flat, brown **Bauhinia forficata**

Leaves evenly pinnate
Leaflets sometimes or always alternate
leaflets narrow, pepper-flavoured, on drooping branches
flowers very small; fruits small, rose-coloured, in showy bunches **Schinus molle**

leaflets large, ovate-lanceolate, odorless; branches spreading
flowers small; fruit globose, up to 2 cm ∅, reddish turning bluish **Pistacia chinensis**

leaflets always opposite
rachis mostly winged
rachis winged or not, leaflets 2 pairs; flowers small, white, in panicles up to 15 cm long; fruit rounded, fleshy, yellowish-green, without appendages **Melicoccus bijugatus**

rachis usually winged, leaflets 4 to 12; flowers small, white, in panicles up to 30 cm long; fruit globose, yellow, with appendages **Sapindus saponaria**

rachis not winged
leaflets few
2(–3) pairs only; flowers blue, in axillary clusters; fruit a small brownish capsule 2 cm ∅ **Guaiacum officinale**

2–4 pairs; flowers pinkish, in terminal panicles; fruit a flat, woody, subcircular pod **Schotia latifolia**

2–5 pairs; flowers inconspicuous, in cylindrical spikes; fruit a flattened indehiscent pod up to 25 cm long **Ceratonia siliqua**

leaflets numerous

8–12 pairs; flowers whitish green, in hanging panicles; capsule woody, ovoid, with winged seeds **Cedrela odorata**

as above but only in juvenile stage, leaves later bipinnate; fruits long pods **Gleditsia triacanthos**

leaflets up to 20 pairs

flowers yellow, in showy 'spikes'; pod narrow, cylindrical **Cassia spectabilis**

flowers cream-coloured to reddish, clusters few-flowered; pods flattened, somewhat constricted **Tamarindus indica**

Leaves bipinnate

Pinnae few, accompanied by spiny stipules

pinnae short, only 2 pairs of leaflets; flowers greenish-white, in rounded heads; pod inflated, somewhat curled; seeds embedded in pulp **Pithecellobium dulce**

pinnae long, pendent, very narrow, with numerous tiny leaflets; flowers yellow, in short racemes; pod about straight, subcylindric, constricted **Parkinsonia aculeata**

pinnae numerous; leaves or branches armed or not

branches and leaves not armed

leaflets tiny; flowerheads yellow, spherical; pods narrow, usually constricted (see also 'phyllodes') **Acacia baileyana, decurrens, dealbata**

leaflets larger

flowers small, cream-coloured, in cylindrical spikes; pods flat, brown, up to 10 cm long **Albizia lophantha**

flowers large, showy, open, red to crimson, in open clusters; pod woody, flattened, reddish-brown, up to 50 cm long **Delonix regia**

flowers roughly bell-shaped, bluish-violet, in large panicles; capsule flat, suborbicular, with winged seeds **Jacaranda mimosifolia**

leaflets rather large

imparipinnate, margin slightly serrate, glands absent; flowers lilac-coloured, in showy panicles; fruit an ovoid yellowish drupe 1 cm ∅ **Melia azedarach**

paripinnate, margin entire, glands present; flowers greenish-white to cream-coloured, with long stamens;

leaves folding up
flower-clusters forming false umbels; pods flat, papery, greyish-yellow **Albizia lebbek**

flower-clusters in open 'heads'; pods flattened, rather hard, blackish **Samanea saman**

branches armed with simple or ramified spines
leaflets many, small; spines simple
flowerheads spherical, yellow to golden yellow; pods narrow, flattened, curved or twisted **Acacia karroo, raddiana** etc.

flowers cream-coloured, in long spikes; pods flattened, almost straight **Prosopis juliflora**

leaflets larger (in juvenile stage leaves pinnate); spines branched; flowers small, yellowish, in dense racemes; pods flat, up to 40 cm long **Gleditsia triacanthos**

DESCRIPTIVE PART

Photo 1. Flamboyant – *Delonix regia*.

MAGNOLIACEAE, the Magnolia family

This family of trees and shrubs consists of 10 to 12 genera with a total of 230 species, and is native in temperate to tropical parts of America and Asia, including SE Asia. The main genus *Magnolia* contains about 80 species of which many are cultivated producing numerous cultivars and even interspecific hybrids. Bean, in a recent publication (1973) describes about 90 species, hybrids, varieties and cultivars grown in Britain, and says that 'perhaps no group of exotic trees gives more distinction to a garden than a comprehensive collection of magnolias.' Indeed the species dealt with below, although it originates in Republican North America, could surely be declared the 'Royal Magnolia'. It is hard to understand why one author once called this species *Magnolia foetida*, even if it is supposed to have an unpleasant smell. – Very few species are grown successfully in the Canary Islands, the most frequent being

Magnolia grandiflora Linné – Southern Magnolia

This large-flowered species sometimes also called the White Laurel Magnolia (or 'Bull Bay') is native in the Southeastern United States but is now grown in parks and gardens of most subtropical and temperate countries all over the world. Usually a tree up to 20 m in height. Trunk erect, short, about 80 cm in diameter or more; bark dark grey, fissured, with large lenticels. Crown dark and dense. Evergreen. Leaves large and shiny, in terminal clusters, short-stalked; blade leathery, oblong-obovate, up to 25 cm long and 10 cm wide; margin entire but frequently somewhat folded. Leaves rusty and downy beneath. Flowers solitary, short-stalked, fragrant, usually white, 12 to 15 cm across. Fruit large, more or less cone-shaped, tomentose, up to 12 cm long; many-seeded.

The species is propagated from seeds and cuttings. In these islands it thrives best between 100 and 500 m above sea level in somewhat moister or cloudier climates. It makes an excellent street tree and an admirable solitary in larger gardens but requires frequent watering. According to Uphof the bark is used as a stimulant tonic and diaphoretic.

Ref.: Bean; Bircher; Chanes; Chittenden; Eliovson; Encke; Graf; Harrison; Kunkel 69; Lombardo 58; Menninger 64; Mitchell; Neal; Uphof.

Fig. 1. *Magnolia grandiflora*; leaf branch, flower and fruit cone = 1/2

Magnolia campbellii Hook.f. & Thoms. – Campbell's Magnolia

Although less frequent than the tree described above, this rose-coloured Magnolia native in the Himalaya (from Assam to Sikkim) appeals with its gorgeous pink flowers 12 to 15 cm across. The species is deciduous and flowers while leafless. According to Bean and other authors it reaches over 30 meters in the wild, however our cultivated specimens are smaller and may be described briefly as follows:

Trunk straight or somewhat bent, bark grey and fissured. Crown open, spreading; lower branches pendent. Leaves oval-shaped or obovate-elliptic, stalked, pointed at apex, 12 to 15(20) cm long, entire or slightly undulate; green and glabrous above, tomentose beneath. Strong and conspicuous nerve system. Petals pink, graduating to crimson towards their centre; fragrant. Fruiting cones narrow and often bent.

The species is best propagated from rooted cuttings. It thrives well at medium altitudes but is susceptible to damage by frost and strong winds. Sometimes a white-flowering variety is also seen.

Ref.: Chittenden; Encke; Graf; Harrison; Kunkel 69; Menninger 62; Mitchell.

A second genus of this family – *Liriodendron* – with its famous Tulip Tree (*L.tulipifera* L.) native in North America, has not been noticed in these islands. It is a large deciduous tree with lobed leaves and bell-shaped, greenish-white or somewhat variegated flowers.

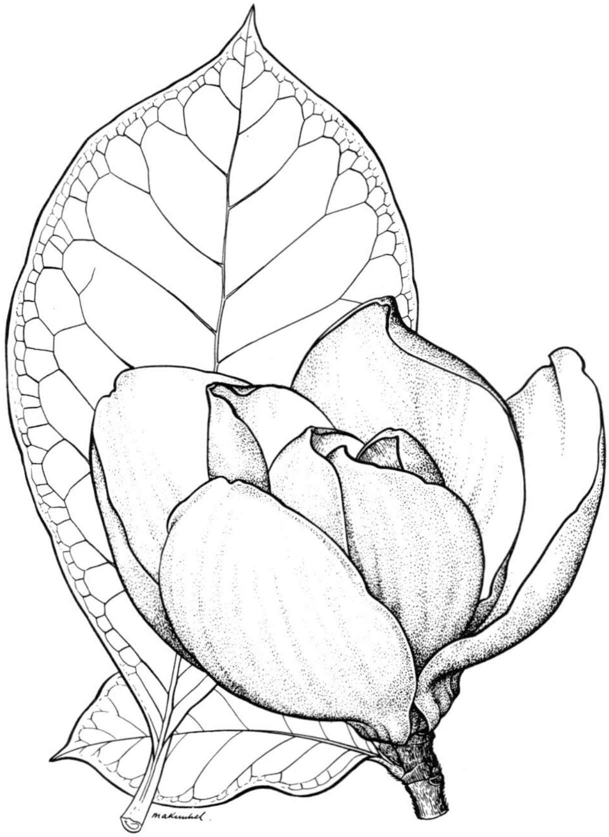

Fig. 2. *Magnolia campbellii*; leaf and flower = 1/2

ANNONACEAE, the Custard-apple family

A large, mainly tropical family of trees and shrubs, according to Willis of approximately 120 genera and 2.100 species. Neal reports more than 70 genera and about 700 species. – In its natural distribution the family is not found in these islands, however some species of *Annona* are cultivated for their edible fruits. The Ylang-Ylang (*Cananga odorata* Hook.f.) is found in the Orotava Botanical Garden, Tenerife. Of certain economic importance is also *Monodora myristica* (Gaertn.) Dunal, the Calabash Nutmeg, apparently not grown in the Canaries.

Annona cherimolia Miller – Cherimoya

A small deciduous tree from western South America much cultivated for its delicious fruits. The tree may reach 6 to 8 m in height; trunk short, bark dark grey, rough or somewhat scaly. Leaves alternate, in a flattened row on spreading or even pendent branches. Stalk short, reddish, and slightly pubescent. Blade ovoid-elliptic, herbaceous, 8 to 12 cm long and 4 to 6 cm wide, minutely tomentose beneath; new leaves somewhat reddish. Flower buds narrow; flowers solitary or in small axillary clusters, fragrant, greenish and rather inconspicuous. Fruit large (up to 15 cm in diameter) and roughly globose, green, fleshy, of a scaly appearance and variable in form depending on the particular variety.

There are many varieties known. The species is propagated from seeds and cuttings, and thrives best in a more subtropical climate; in the tropics higher elevations are recommended for this otherwise undemanding fruit tree. – Martinez reports that pulverized seeds, which are poisonous, are used as an insecticide.

Ref.: Adams, Barrett 56; Bircher; Burkill; Calabria; Chittenden; Degener; Encke; Kunkel 69; Martinez; Molesworth Allen; Mortensen & Bullard; Neal; Pesman; Purseglove; Soukup; Uphof.

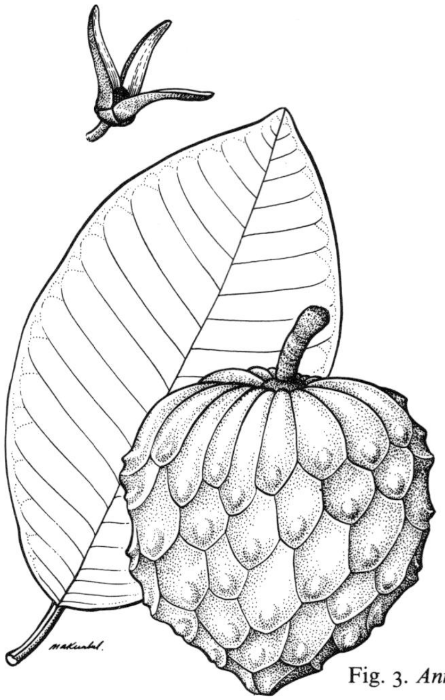

Fig. 3. *Annona cherimolia*; leaf, fruit and solitary flower = 1/2

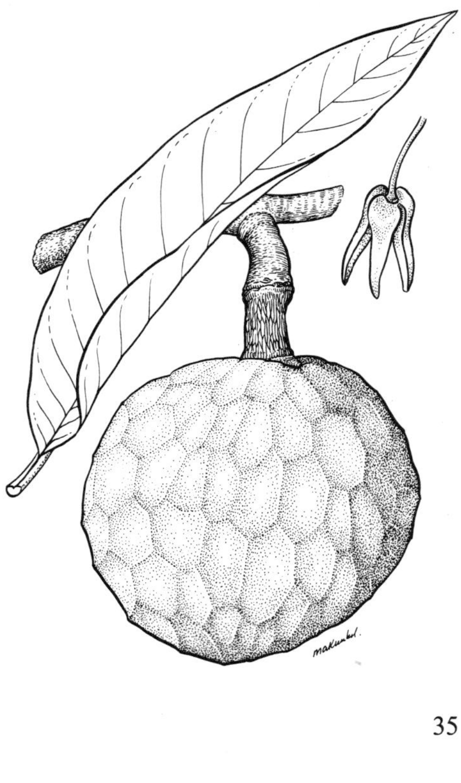

Fig. 4. *Annona reticulata*; details = 1/2

Annona muricata Linné – Soursop

The 'Guanábana tree' of the American Indians is equally small but is preferred for true tropical climates. It is distinguished from the above by its brownish and smooth bark, the evergreen foliage, the more globose flower-buds, the lanceolate or oblong glossy leaves up to 15 cm long, and its roughly kidney-shaped, spiny fruits.

The Soursop is native in the Caribbean region. It is propagated from seeds. Flowers and leaves have a peculiar, rather unpleasant smell. It is grown for its fruits which are eaten raw or as a preserve, in ice-creams and refreshing drinks. The fruit is anti-scorbutic, and the seeds are astringent (Burkill). Leaves are used for skin diseases and rheumatic complaints; Irvine reports that leaves are used in Gambia 'to get rid of bed bugs'.

Ref.: Adams; Barrett 56; Bircher; Burkill; Calabria; Chittenden; Cobley; Encke; Irvine; Little & Wadsworth; Martinez; Molesworth Allen; Mortensen & Bullard; Neal; Purseglove; Uphof.

There are two more species grown in these islands, although less frequent than the above cited: the Custard-apple (*Annona reticulata* L.), and the Anón or Sugar-apple (*A.squamosa* L.). Both species are fruit trees like the aforementioned. A hybrid between *Annona cherimolia* and *A.squamosa* is known as Atemoya, and is said to be a very promising cross.

Fig. 5. *Annona muricata*; drawing = 1/2

LAURACEAE, the Laurel family

A family of trees and shrubs of tropical and subtropical forests, with main distribution centres in America and Southeast Asia. Rendle believes the Lauraceae to consist of 40 genera with about 1.000 species, whereas the 'Willis Dictionary' cite the number of 32 genera with 2.000 to 2.500 species (Purseglove: 47 genera, 1.900 spp.). An aberrant genus – *Cassytha* – of *Cuscuta*-like parasites might be rightfully placed in a distinctive family: the Cassythaceae. – The Lauraceae has produced many useful timber trees. Species of *Apollonias*, *Laurus*, *Ocotea* and *Persea* are natives of the Canary Islands and main components of the indigenous Laurisilva. The Mediterranean *Laurus nobilis* L. (Laurel Bay) is grown in a few gardens for its aromatic leaves.

Cinnamomum camphora (L.) Presl. – Camphor Tree

A medium-sized tree (15–20 m) from East Asia, widely cultivated for its medicinal properties. Trunk high, up to 1 m in diameter, base often with buttresses; bark brownish-grey, thick and almost corky, deeply fissured or scaly. Wide spreading dense crown. Evergreen. Leaves alternate, long-stalked; blade ovate-elliptic, acuminate, more or less herbaceous, glossy green, 8 to 10 cm long and 3 to 5 cm wide, with pronounced lateral nerves and glands at their base. Leaves with a strong camphor smell. Flowers small, greenish-white, in open terminal or axillary panicles. Fruit roughly olive-shaped, almost black, up to 1 cm in diameter.
Propagation from seeds and cuttings, grown best at lower elevations. According to Barrett the Camphor Tree 'grows well in salt air' and is used as a windbreak for *Citrus* plantations. The wood is used for furniture and general construction, especially for 'chests because the odor is repellent to insects' (Little & al.). Most important (until synthetic camphor was invented) was the volatile camphor oil distilled from wood, roots and leaves and employed for medicinal (narcotic) purposes; internally taken it is proved to be poisonous. Menninger (1967: 226) reports at length on the virtues of and the magic connected with the Camphor Tree. – The species is sometimes cited as *C. camphora* (L.) Nees & Eberm., or (L.) Siebold.

Ref.: Adams; Barrett 56; Bean; Bircher; Burkill; Chittenden; Corner; Encke; Harrison; Irvine; Kunkel 69; Lombardo 58; Long & Lakela; Menninger 64, 67; Neal; Purseglove; Uphof.

Fig. 6. *Cinnamomum camphora*; flowering branch and fruit = 1/2, solitary flower = natural size.

Cinnamomum zeylanicum Blume – Cinnamon Tree

A well-known and widely cultivated South Asian tree which is claimed by several authors (Breyne, Blume, Nees); originally described by Linné ('Species Plantarum' 1753: 369) as *Laurus Cinnamomum* and harvested since time immemorial, long before science was 'invented'. It's just a tree of which the young bark produces a delicate spice but which – in the 16th Century – nearly started a war between the Dutch and the Portuguese, and later between the British and both of them! (Burkill). The species is now planted in many tropical and subtropical countries; it makes a beautiful shade tree in the Canary Islands and other places where its commercial value is of no importance.
Tree 8 to 12 m in height, with a dense wide spreading crown. Trunk short, stout or somewhat bulgy, with a considerable rootsystem; bark greyish-brown, scaly, or corky in older specimens. Terminal branches pendent. Foliage evergreen. Leaves alternate or subopposite, petiolate, glossy pale green, ovate-lanceolate, 7 to 10 cm long and 2,5 to 4 cm wide; strongly 3-nerved. Young leaves purplish and shiny. Flowers small, cream-coloured, in axillary racemes. Fruit more or less olive-shaped as in the previously described species but of a more bluish-black colour.

The Cinnamon Tree, if cultivated commercially, is kept shrubby because its branchlets and new shoots are frequently pruned and the thin bark pealed off. The dried bark makes an important spice, the familiar cinnamon. Leaves, bark, trunk and roots yield several essential oils which are used as extracts, and in medicine as an astringent and antipyretic. The species is propagated from seeds which should be sown soon after harvesting because they have a very limited viability (Purseglove).

Ref.: Adams; Bircher; Burkill; Chittenden; Cobley; Corner; Encke; Esdorn & Pirson; Irvine; Kunkel 69; Little & al.; Lombardo 58; Neal; Purseglove; Uphof.

Fig. 7. *Cinnamomum zeylanicum*; drawings = 1/2

Persea americana Miller – Avocado Tree

Probably the most commonly cultivated and most widely distributed member of the Laurel family, said to be native in Central America, now in almost every tropical and subtropical region. It produces one of the most delicious and nutritious fruits which really should be treated as a 'vegetable'. Purseglove calls it 'the most nutritious of all fruits'. In my opinion, once accustomed to the taste of the avocado pear – if it were cheap enough and always available – one could easily become a voluntary vegetarian …
Although more recently created varieties are kept as low-growing shrubby trees, the 'primitive' Avocado Tree may reach 20 m in height, with a pronounced trunk up to 70 cm in diameter. Bark grey or greyish-brown, rough or fissured. Crown more or less pyramidal or spreading. Foliage usually deciduous although some cultivars in favourable climates tend to have semi-persistant leaves. Twigs green, shiny, in some forms somewhat angular, with mucilage-glands. Leaves alternate, long-stalked; stalks up to 4 cm long. Blade oblong-elliptic, with a cuneate base and a long-pointed apex, herbaceous, dark green, 10 to 20 cm long; the size and shape depending on the variety grown. Flowers yellowish-green, in large, usually terminal panicles. Fruits rounded, pear-shaped or almost melon-like, with a smooth or rough skin, a buttery pulp, and a large, rounded or egg-shaped seed.
The growth habit, leaf form, fruit and seed, all depend on the particular variety or cultivar. Fruits might have a thin, smooth green skin or be black, thick-skinned and warty, and their weight may vary from about a hundred grammes to over 3 kg. The fruit flesh is rich in vitamins, fat and proteins; it is eaten raw (with vinegar, oil, pepper and salt) or in salads. The rind expels intestinal parasites, and 'leaves and bark are used in home remedies for their stomachic, pectoral, emmenagogue, antiperiodic and resolutive properties' (Uphof). Propagation usually from seeds which take a considerable time to germinate; commercially important cultivars are grafted on common stock. Seeds are often seen half-sunken in water and after germination may produce an admirable house plant, an experiment which requires much patience. Avocado trees prefer a warm humid climate, are susceptible to frost and strong wind, and may be attacked by a deadly enemy: *Phytophthora cinnamomum*, a root-fungus of badly drained soils. In the Canary Islands grafting of *Persea americana* on stocks of the native *P.indica* was promising but the afore mentioned killer-fungus seems to work faster. Some cultivars are grown in even fairly saline soils. – Commonly cited synonym: *Persea gratissima* Gaertn.; alternative common name: Alligator Pear.

Ref.: Adams; Barrett 56; Bircher; Burkill; Calabria; (Chittenden); Cobley; Corner; Encke; Esdorn & Pirson; Irvine; Keay & al.; Little & Wadsworth; Lombardo 58; Long & Lakela; Malo; Martinez; Moeller 71; Molesworth Allen; Mortensen & Bullard; Neal; Pesman; Purseglove; Soukup; Uphof.

Fig. 8. *Persea americana*; drawing = 1/2

MENISPERMACEAE, the Moonseed family

A mainly tropical family of trees, shrubs, herbs and vines which consists of 65 to 70 genera with about 350 species. In these islands only

Cocculus laurifolius DC. – Laurel-leaved Moonseed

According to Uphof a 'woody vine', however it is a large shrub to medium-sized tree (7–10 m) in Canary gardens, with a wide-spreading crown. Trunk short, or branched at the base forming several trunks which reach up to 40 cm in diameter. Bark dark grey, somewhat warty. Foliage evergreen; terminal branches knotty and tending to droop. Leaves alternate, petiolate, oblong-lanceolate with an elongated tip; blade glossy green, herbaceous, 10 to 15(18) cm long and 3 to 5 cm wide, strongly 3-nerved. Flowers very small, pale green, axillary, in clustered or subcylindric panicles. Fruits small, spherical, nearly black when ripe, 3 to 4 mm in diameter.

An Asiatic (Himalayan) species, frequently cultivated in Italy and in the south of France for its picturesque dense and spreading crown. It makes a good shade tree. Propagation from seeds and cuttings. The bark is 'source of alcaloids, cocculine and coclaurine, having properties of curare' (Uphof).

Ref.: Bean; Burkill; Chittenden; Encke; Kunkel 69; Lombardo 61; Uphof.

Fig. 9. *Cocculus laurifolius*; flowering branch and fruits = 1/2

PLATANACEAE, the Plane-tree family

A monogeneric family of northern temperate regions, with about 10 species and a 'troublesome' but very successful hybrid which is widely cultivated, and described as follows:

Platanus × *hybrida* Brot. – London Plane

Also known as Spanish Plane, and usually cited as *Platanus* × *acerifolia* (Ait.) Willd., or *P.* × *hispanica* Muenchh. A rather variable hybrid between *Platanus orientalis* L. and *P.occidentalis* L. which deserves more investigation as much material from the Mediterranean zone and perhaps all specimens cultivated in the Canary Islands differ in leaf-shape from the typical 'London Plane'; this material could belong to *P.* × *hispanica* if this taxon is recognized as a distinct and different hybrid (i.e. Chittenden, Ruiz de la Torre). Here we follow the treatment presented in 'Flora Europaea'. For the history of cultivated Plane trees see the thorough treatment by Elwes & Henry.
Large tree up to 40 m tall; when planted at roadsides often heavily pruned and much smaller. Trunk usually straight, 60 to 80 cm (1,5 m) in diameter; bark grey, scaling off in flakes to expose the paler young bark. Crown high, lower branches spreading. Foliage deciduous. Leaves alternate, long-stalked; blade more or less triangular, usually 5-lobed, coarsely toothed, subcoriaceous, shiny green above; young leaves pubescent. Flowers small but numerous, in perfectly spherical heads which develop into fruits without changing their form; ripe fruiting heads up to 4 cm in diameter, usually two on an elongated pendent stalk.
Propagation from woody cuttings. An excellent roadside and avenue tree; also in larger parks and gardens.

Ref.: (Chanes); (Chittenden); (Elwes & Henry); (Encke); (Fitschen); (Harrison); (Hart & Raymond); (Kunkel 69); (Lombardo 58); (Milano & Molinari); (Mitchell); Polunin & Huxley; Polunin & Smythies; Ruiz de la Torre; Tutin 64; (Willis).

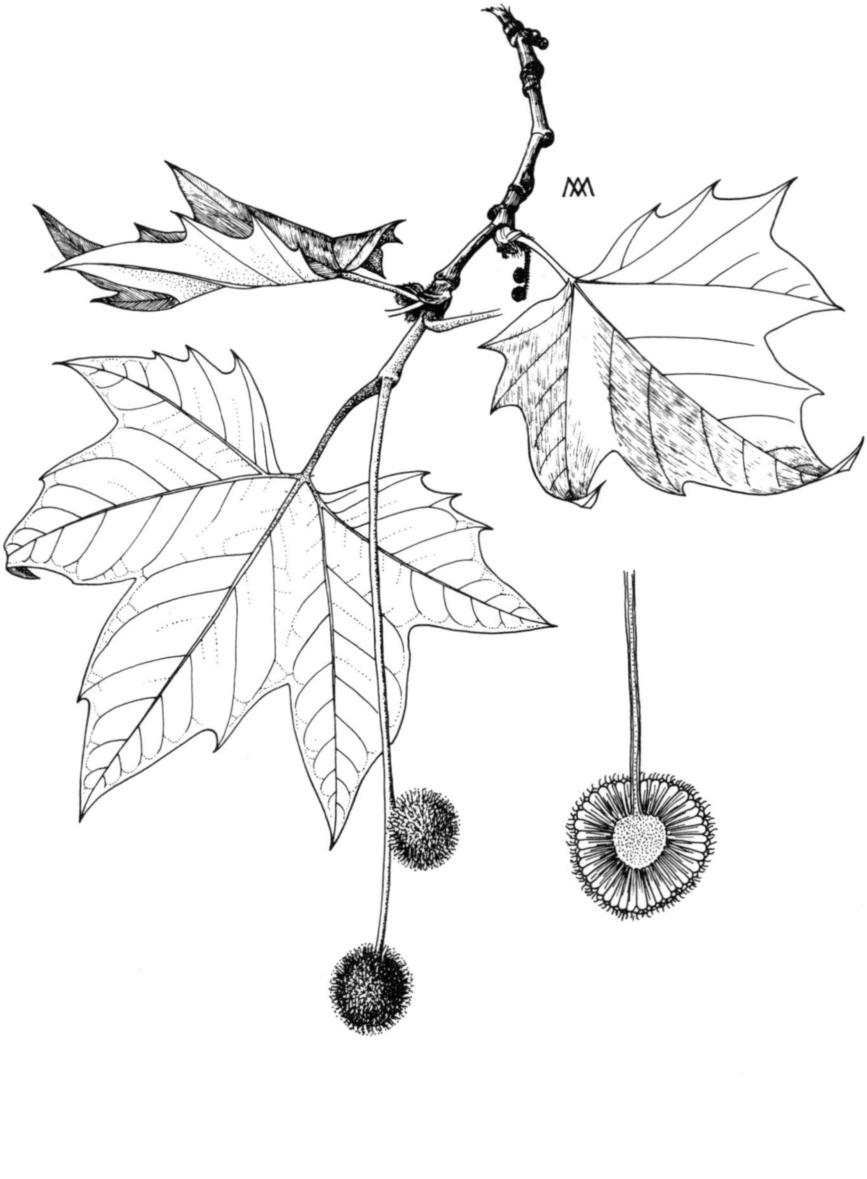

Fig. 10. *Platanus × hybrida*; drawings = 2/3 reproduced from Kunkel: 'Arboles exóticos'

ULMACEAE, the Elm family

The Elm family (if we include *Celtis*, *Trema*, *Zelkowa* etc.) consists of 13 to 15 genera with 150 to 200 species of trees and shrubs which are found from the tropics to temperate zones. The genera *Celtis* and *Trema* are frequent components of tropical forests, whereas *Ulmus* seems to be restricted to more temperate climates. Both species described below are naturalized in the Canary Islands.

Celtis australis Linné – European Hackberry

Sometimes called the 'Netteltree', this deciuous species is a native of the Mediterranean region and the Near East. About 15 to 20 m tall, with a short but pronounced trunk up to 80 cm in diameter; bark dark grey, smooth or scaly, with horizontal grooves on the branches. Crown spreading and generally rounded; new branchlets slightly tomentose. Leaves alternate, petiolate, more or less glaucous, often pendent; blade ovoid-oblong with an unequal base, herbaceous to subcoriaceous, rather rough above and somewhat downy beneath, up to 12 cm long; margin strongly dentate. Flowers small, stalked, greenish, axillary, solitary or in groups of 2 or 3. Fruits globose drupes up to 1 cm in diameter, purplish or black, 1-seeded; the fruits are said to be edible.
The Hackberry is propagated from seeds and cuttings. A quite hardy species which resists a drought as well as an occasional frost. The fairly hard wood is used for construction and manufacturing small implements. Leaves and fruits are astringent, antiperiodic, and used for dysentery (Font Quer); leaves are consumed by animals. The species is often planted as a roadside tree especially in towns; it becomes a 'weed' on irrigated land.

Ref.: Bean; Bircher; Chanes; Chittenden; Encke; Fitschen; Font Quer; Harrison; Kunkel 69; Lombardo 58; Mitchell; Polunin & Smythies; Ruiz de la Torre; Tutin 64; Uphof.

Fig. 11. *Celtis australis*; fruiting branch = 1/2

Ulmus minor Miller – Elm Tree

Considering its numerous synonyms we deal, obviously, with a much confused species native or established in most European countries, parts of North Africa and extending as far as Asia Minor. Sometimes cited as *Ulmus campestris* auct. non L., *U.carpinifolia* Suckow, *U.glabra* Miller, non Hudson, etc. – The author hopes that the tree described below corresponds to the name given above.

Tree up to 20 (25) meters tall, its basal and root-shoots have a tendency to invade surrounding land. Trunk erect; bark dark grey and slightly fissured turning very dark and furrowed in older specimens. Crown more or less pyramidal; young branches somewhat tomentose and rather flexible. Deciduous. Leaves alternate, short-stalked; blade ovate to obovate, unequal at base, dark green, herbaceous but with a rough surface, up to 8 cm long and 3 to 5 cm wide; margin dentate-serrate. Flowers greenish, in short dense clusters on previous year's growth. Fruit winged, one-seeded, green turning papery and yellow, up to 1,5 cm across.

Propagation from seeds, cuttings and suckers; cut trees usually produce strong new growth. The species is often planted as a roadside tree at somewhat higher altitudes. According to Ruiz de la Torre specimens are known to reach 40 m in height, with a diameter of 2,70 meters. The hard and durable wood is used in carpentry and construction, for charcoal, fuel etc. The bark is astringent and is also used for skin diseases. – A few specimens of the broad-leaved American White Elm (*Ulmus americana* Mill.) are also grown in these islands.

Ref.: (Chanes); (Chittenden); (Encke); (Fitschen); (Font Quer); (Lombardo 58?); (Mitchell); Ruiz de la Torre; Tutin 64; (Uphof).

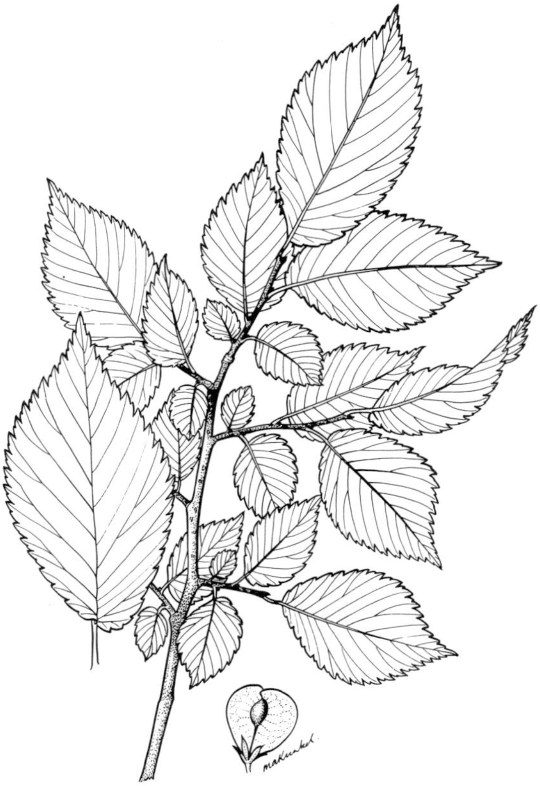

Fig. 12. *Ulmus minor*; branch, and solitary flower = 1/2

MORACEAE, the Fig family

This rather complex family consists of 55 to 60 genera and 1.200 to 1.400 species most of which are trees or shrubs, a few are vines or herbs. The main distribution centre of the Moraceae is found in the tropical zones of both hemispheres. Some species are known to be very poisonous (i.e. *Antiaris toxicaria* Lesch); the latex of *Castilla elastica* Cerv. yields caoutchouc, *Brosinum galactodendron* D.Don is the Cow- or Milk-tree the milky latex of which is used like ordinary milk, and species of *Dorstenia* are cultivated for their curious flower heads. – According to modern taxonomy *Cannabis* and *Humulus* are excluded and form the segregated family of the Cannabidaceae. Most Moraceae belong to the genus *Ficus* which is rather troublesome in every sense and in eternal need of revision. According to Neal and to 'Willis Dictionary' the genus consists of some 800 species; Ehrendorfer and Rendle consider some 700 species for the genus *Ficus*, whereas Encke gives the incredible number of almost 2.000 species. For further discussion on the species-number see also Condit (p. 4). – Whilst referring to the previously mentioned word 'troublesome' I mean it, not only in the numeric or classification sense: figs often grow in walls bringing down entire constructions, others are true killers which survive after strangling their initial host. On the other hand some fig species are sacred as we shall see in the following text, some produce edible fruits and some others are favoured shade trees. This work deals with only 18 of the 20 or more species of fig trees found in Canary gardens.

Ficus afzelii G.Don ex Loudon – Loquat-leaved Fig

This species native in West Tropical Africa is better known under its synonym *Ficus eriobotryoides* Kth. & Bouché, a name very appropriate because of the loquat-like leaves. However: priority *has* priority, even in figs, and we have to accept *F.afzelii* as the valid name. – An alternative common name is 'Old Calabar fig' (Condit).

Although this species in its native habitat may begin its life cycle as an epiphyte, specimens in gardens (propagated from cuttings) usually behave like any 'normal tree'. They reach a size of 8 to 10 meters only, but in Africa,

Fig. 13. *Ficus afzelii*; fruiting branch and solitary fruits = 1/2

according to Keay & al. '40–80 feet high and up to 7 feet in girth'. Trunk slender. Bark greyish and furrowed. Branches spreading forming a rather flat crown; young branches somewhat rusty, with leaf-scars. Foliage evergreen. Leaves stalked, terminally clustered; petiole hairy. Blade leathery, entire, oblanceolate, with an acuminate or emarginate apex; up to 30 cm long and 8 cm wide. Terminal bud 2,5 to 3 cm long, and silky. Figs sessile, subglobose, very woolly, brownish, up to 3 cm in diameter; said to be edible.
A rare species in these islands growing best at lower altitudes. It is a slow growing tree which has to be treated with tender care. For varieties, discussions of the valid name and for alleged synonyms see Barrett (1948).

Ref.: Aubreville; Barrett 48; (Bircher); Condit; Irvine; Keay & al.

Ficus altissima Blume – Lofty Fig

A very showy tree with a wide spreading crown, native in Southeast Asia. Up to 25 m tall; trunk short, bent, with aerial roots; bark grey, smooth, with lenticels. Probably evergreen in the tropics. Leaves pale green, long-stalked, oblong to ovate-elliptic, glabrous, subcoriaceous, entire but often slightly undulate, up to 30 cm long and 15 cm wide; apex abruptly acuminate, base cuneate. Terminal bud 2,5–3 cm long, slightly tomentose. Figs subglobose, sessile, usually in pairs, orange-coloured when ripe, up to 2,5 cm in diameter; when young covered by fleshy bracts.
The Lofty Fig makes a good 'solitary' in larger gardens; sometimes planted as a roadside tree but it needs much space. Additional names as given by Barrett: Council tree, and High Rubber tree.

Ref.: Barrett 56; Bircher; Condit; Burkill; Kunkel 69.

Ficus aspera Forster – Clown Fig

Usually cited under its synonym *F.parcelli* Veitch ex Cogn. & March., this large shrub or small tree is native in Polynesia. It is a curious member of the genus because this special cultivar has very showy, variegated soft leaves and is much appreciated by collectors of garden plants. Twigs finely pubescent. Leaves unequal at the base, up to 15 (25) cm long and 7 cm wide. Figs single or in pairs, up to 3 cm in diameter often appearing even on the trunk itself (cauliflorous). – Alternative name: Mosaic Fig.

Ref.: (Chittenden); Condit; (Encke); (Graf); (Kunkel 69); Synge.

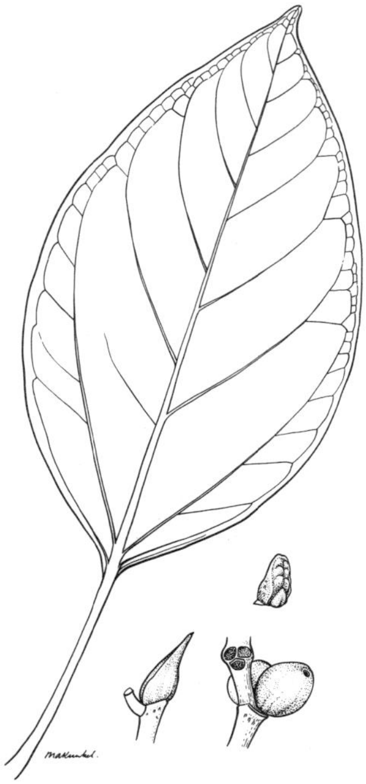

Fig. 14. *Ficus altissima*; leaf and details = 2/5

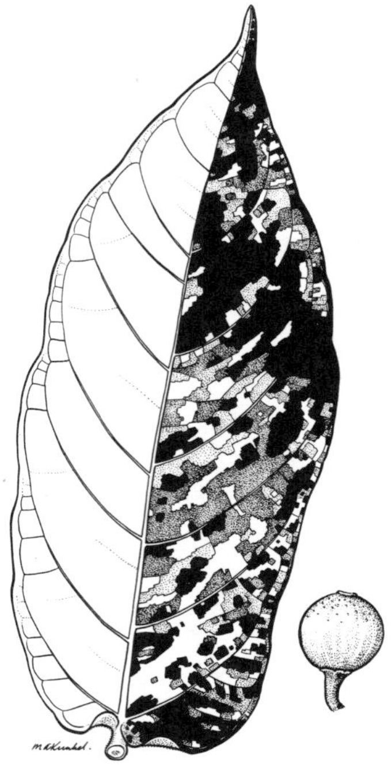

Fig. 15. *Ficus aspera*; leaf and fruit = 2/5

Ficus aurea Nuttal – Strangler Fig

Also known as the Florida Strangling Fig or, more pleasantly, the 'Golden Fig', it appears as a common epiphyte in Southern Florida, its home country. Our garden specimens, however, are grown from rooted cuttings and no germination from seeds was ever observed. – I. J. Condit describes a 'tree-mendous oddity': the Strangler Fig rooted on a *Phoenix* palm.

The tree may reach 12 to 15 m in height and is said to have aerial 'strangling' roots and a somewhat buttressed trunk the bark of which is dark grey and little fissured. Foliage evergreen. The leathery leaves have stalks up to 3 cm long, are ovate-elliptical in outline, up to 10 cm long and 6 cm wide. Margin entire, blade dark green and glossy above and somewhat rusty beneath. Terminal bud more or less conical, 3 to 5 cm long. Young branches green, with elongated lenticels. Fruits axillary, usually in pairs, almost sessile, subglobose, yellowish or spotted, up to 1,5 cm in diameter.

The Strangler Fig makes a showy ornamental in parks and gardens, sometimes also planted along roadsides. It grows best at lower altitudes, and is susceptible to frost.

Ref.: Adams; Chittenden; Condit; Graf; Long & Lakela.

Fig. 16. *Ficus aurea*; branch and details = 1/2

Ficus benghalensis Linné – Banyan Fig

A well-known species suitable for large gardens only; in more tropical climates single specimens may grow into a 'vast forest'. Specimens soon develop aerial roots from the lower branches which, after reaching the ground, transform into trunks supporting the enormous weight of these branches. The famous Great Banyan of the Calcutta Botanical Gardens measured almost 1.000 feet in circumference and had over 600 strong aerial roots or supporting 'trunks'. E. A. Menninger, the admired 'Flowering Tree Man' and perhaps a slow walker, reports the Calcutta tree as 'a banyan tree so enormous and with so many trunks that it takes ten minutes to walk around it'. – This Indian Banyan or Vada Tree (alternative names) is sacred to Hindus. It is native in India, and is frequently cited in respective literature. The following description refers to specimens grown in these islands where the Banyan hardly reaches 20 m in height. Trunk short and stout; bark pale grey, with lenticels and horizontal grooves. Evergreen. Leaves long-stalked, downy when young, later dark green and glabrous above, usually somewhat tomentose beneath; short-stipuled. Blade elliptical, ovoid or even roughly orbicular, leathery, 12 to 15 cm broad; margin entire. Terminal bud short, 2 to 2,5 cm long, tomentose. Figs globose, usually in pairs, reddish-tomentose, sessile, up to 2 cm in diameter.

As the Banyan Fig is said to be sacred its wood is only seldom (and carefully) used for boxes and pannels, and the aerial roots for tent poles. The latex contains a low percentage of caoutchouc but 'thickens into a kind of bird-lime' (Burkill). Fruits are eaten by birds, and also by humans in times of famine. Branchlets and leaves are consumed by cattle and elephants (Barrett). In the Canary Islands, unfortunately the leaves of this species are often infested by a small but rather messy 'white fly'. – The near-related Krishna Fig (*Ficus krishnae*) is treated by Condit as a variety of the Indian Banyan.

Ref.: Barrett 56; Bircher; Burkill; Chittenden; Condit; Corner; El Hadidi & Boulos; Encke; Graf; Huxley; Kunkel 69; Lombardo 58; Menninger 64, 67; Neal; Purseglove; Uphof.

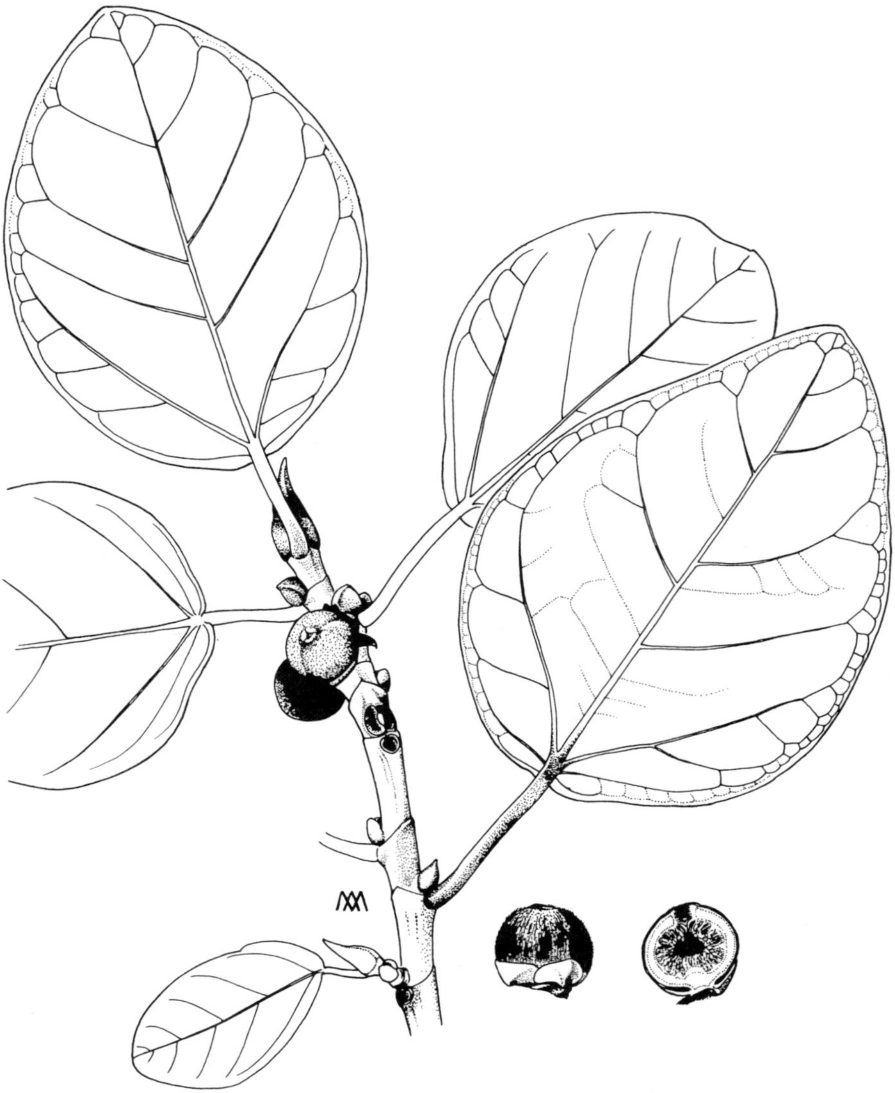

Fig. 17. *Ficus benghalensis*; drawing = 1/2 from Kunkel: 'Arboles exóticos'

Ficus benjamina Linné – Benjamin Fig

Another large Indian tree but also a favoured pot-plant in houses of the temperate zone. It is the Weeping fig, Weeping laurel, Willow fig, Chinese banyan, Java willow and what else of other authors, and placed within the group of 'strangling figs', by Corner. The species is easily recognized by its willowlike drooping branches and the gracious, finely veined leaves. Aerial roots rarely occur, and 'this species seems to have a hard time in deciding whether or not to be a banyan' (Barrett 1956). However Condit says the Florida specimens do have aerial roots, 'some of which formed multiple trunks'. For a generalized description one may summarize the following: Tree up to 20 (25) m tall, with or without aerial roots. Trunk short, in older specimens forming buttresses. Bark grey, smooth or warty. Branches wide-spreading, slender, finally drooping. Foliage evergreen. Terminal bud narrow, up to 1,5 cm long, sharp pointed. Leaves petiolate, ovate or oblong-elliptic, abruptly pointed, somewhat leathery, glossy green, with numerous, closely-spaced fine lateral veins; up to 10 cm long and 4 to 5 cm wide. Figs axillary, sessile, solitary or in pairs, more or less globose, scarlet and spotty, about 1 cm in diameter. – This description may apply to the typical variety; several other varieties are described by Barrett 51, Condit, and others.

Corner believes the species to be native in the Indo-Malayan region as such. Its timber seems little durable. Strips of the tough bark 'made into rope possess a fair degree of tenacity' (Burkill). Fruits apparently not eaten. – The Benjamin Fig will not stand up to frosty climates nor to very arid conditions.

Ref.: Adams; Barrett 51, 56; Bircher; Burkill; Chittenden; Condit; Corner; Encke; Graf; Harrison; Little & al.; Neal; Purseglove.

Fig. 18. *Ficus benjamina*; branch and details = 1/2

Ficus carica Linné – Common Fig

The most widely cultivated species of which not less than 700 (!) varieties have been described (Condit). A deciduous tree, quite hardy in cooler climates, of historical fame and planted in most countries with a Mediterranean climate. Probably native in the Near and Middle East, although Tutin thinks it 'perhaps native in the southern parts of the Iberian peninsula, Italy, the Balkan peninsula and U.S.S.R.'
A small tree (6 to 8 m) with a short and twisted trunk, frequently with basal shoots. Bark pale grey; crown open with rather irregular branches. Terminal bud short and stout. Leaves hard-herbaceous, with a rough surface and of variable size and shape; long-stalked. In most cultivated forms usually 3 to 5-lobed; blade between 10 and 25 cm long, dark green, with a pronounced nerve system. Figs mostly solitary, axillary, green (or yellow, brown, purplish or even black), more or less pear- or drop-shaped, with either a velvety or glabrous skin; in certain varieties up to 6 cm in diameter but normally of more modest size.
The species is mostly grown for its edible fruits; in fact it is one of the oldest cultivated (or wild-harvested) woody plants of the world and appears in songs and legends of historial and mythological background. Trees are reported to live for over 200 years. The fruits are eaten raw, dried or otherwise preserved; 'when dried and ground used as substitute for coffee' (Uphof). Figs are high in calcium, sugar, iron and copper (Mortensen & Bullard).
Leaves are eaten by cattle and, according to the Bible, were used as the first kind of mini-dress by Adam and Eve. In some places the latex is used to coagulate milk. Various medicinal properties are cited by Martinez; its application for boils and other skin infections is reported by Polunin & Huxley and, especially, by Font Quer who describes the chemical composition of the fruit. Fruits are mildly laxative. The world production of fresh figs (1968) is given by Schütt at about 1.255.000 tons.

Ref.: Barrett 56; Bean; Beyron; Bircher; Burkill; Chanes; Chittenden; Condit; Encke; Esdorn & Pirson; Fitschen; Font Quer; Graf; Kennard & Winters; Kunkel 69; Long & Lakela; Martinez; Mitchell; Mortensen & Bullard; Neal; Pesman; Polunin & Huxley; Polunin & Smythies; Purseglove; Rauh; Ruiz de la Torre; Schaeffer; Schütt; Tutin 64; Uphof.

Fig. 19. *Ficus carica*; branch with fruit = 1/2

Ficus elastica Roxb. ex Hornem. – India Rubber Tree

Besides the Common Fig probably the most frequently cultivated species and (with its broad-leaved cv. 'decora') certainly the most common ornamental fig to be grown in pots and flower boxes. The species is a native of Northern India, whereas the popular form 'decora' is said to have been raised by a Belgian nurseryman (Condit).

The primitive form (var. *elastica*) may be described as follows: Tree up to 30 m in height, in these islands usually smaller (8–12 m). Trunk short, somewhat irregular due to early forming buttresses; in older specimens aerial roots may appear. Bark grey, with lenticels and shallow horizontal grooves. Crown spreading, terminal branches often pendent. Evergreen. Leaves alternate, long-stalked, entire, ovate or oblong-elliptic, leathery, shiny green above, up to 30 cm long and 12 cm wide, with a very fine venation. Terminal bud 10 to 25 cm long. Figs axillary, usually in pairs, more or less cylindrical, 1 to 3 cm long, greenish with darker spots. Leaves of the cv. 'decora' are much larger and glossy dark green; less frequent in cultivation is the spotted-leaved cv. 'Variegata'.

Condit, Barrett and other authors report that the India Rubber Tree may reach up to 60 m in height; Menninger (67) says that this species also 'develops pillar roots in the wild and becomes an enormous banyan'. According to Burkill many thousands of acres of *Ficus elastica* were planted in India and Malaysia and cultivated until the Para Rubber Tree (*Hevea brasiliensis*) was introduced. – The species is usually propagated from either single cuttings or branches prepared in the 'marcottage system' (or 'mossing'). Alternative common names: Rubber Fig, India Rubber Fig, and Snake Tree. Several authors cite also the name 'Berlin Weed' because the species is very common in German apartment houses.

Ref.: Barrett 51, 56; Bircher; Burkill; Chanes; Chittenden; Cobley; Condit; Corner; El Hadidi & Boulos; Encke; Graf; Harrison; Irvine; Kunkel 69; Little & al.; Lombardo 58; Long & Lakela; Menninger 67; Neal; Pesman; Purseglove; Uphof.

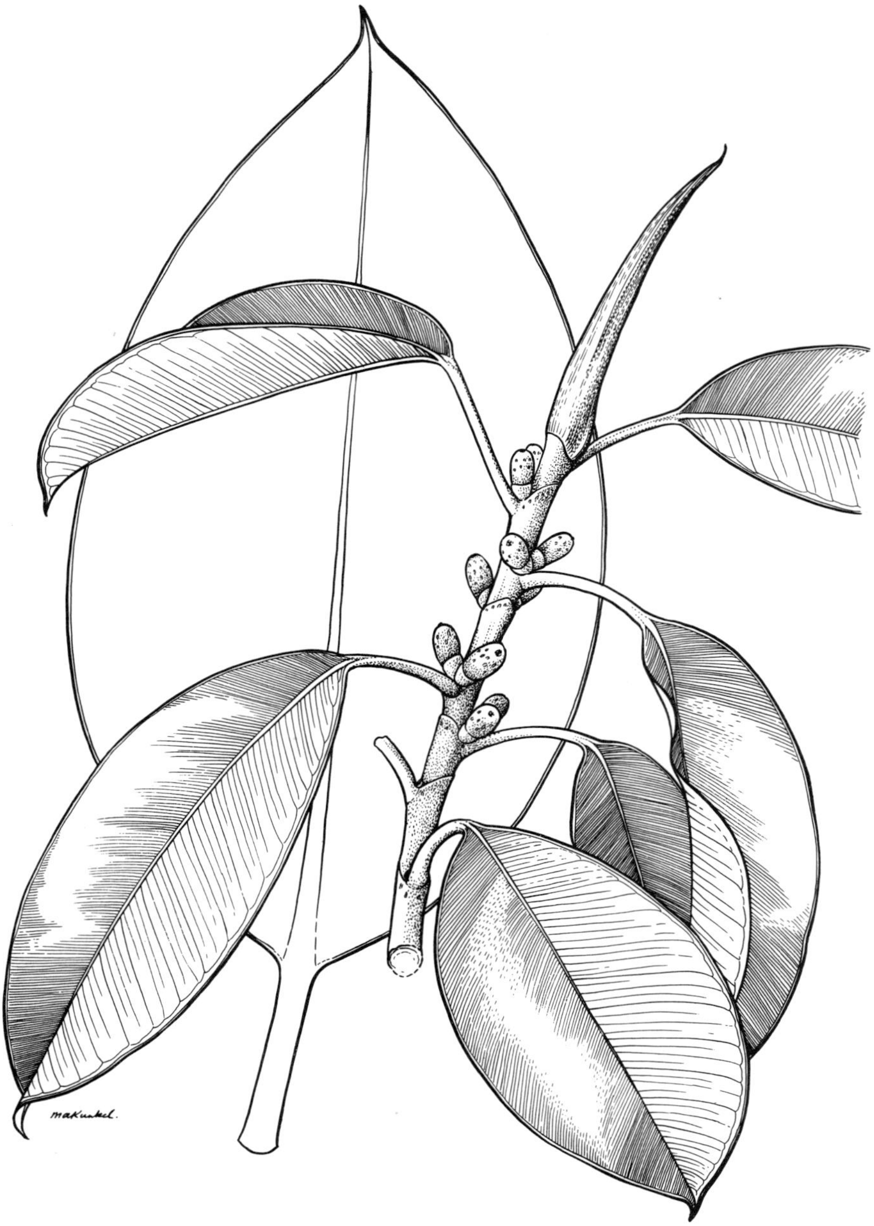

Fig. 20. *Ficus elastica*; drawing = 1/2

Ficus jimenezii Standl.

This is a larger tree, only briefly deciduous and which somewhat resembles *Ficus virens* except that the terminal branchlets are of angular appearance and the leaves are always rather distant. Terminal bud somewhat curved. Figs sessile. Leaves up to 15 cm long and 8 cm wide. The species is probably of American origin and rather rare in Canary gardens.

Fig. 21. *Ficus jimenezii*; branch and detail = 2/5

Ficus lyrata Warburg – Fiddle Fig

Also known as Fiddle-leaf Fig, Banjo Fig, and Lyrate-leaved Fig, this West African species is sometimes cited as *F.pandurata* Sander (non Hance). A tree 12 to 15 m high in cultivation but in the wild said to reach up to 25 meters (Barrett 56). Trunk short, 50 to 60 cm in diameter; bark dark grey, very rough or fissured. Crown spreading and very dense. Evergreen. Leaves very large up to or over 30 cm long, fiddle-shaped, coriaceous, glossy green, with strong lateral veins, short-stalked and more or less clustered in terminal position. Branchlets usually covered with persistant stipules. Figs sessile, single or in pairs, globose, fleshy, purplish with pale dots, and with a velvety skin; up to 5 cm in diameter, and edible.

Although this species (like most figs) is said to begin its life-cycle as an epiphyte, propagation of garden plants is done from cuttings rooted on the tree (mossing). The Fiddle Fig makes a good garden tree and grows best at lower altitudes in warm, moist climates. It is also found as an indoor pot-plant in temperate zones.

Ref.: Aubreville; Barrett 48, 56; (Chittenden); Condit; Encke; Graf; Irvine; Keay & al.; Kunkel 69; Little & al.; Moeller 71; Neal; Synge.

Ficus macrophylla Desf. ex Pers. – Moreton Bay Fig

A large, evergreen Australian tree which, according to Barrett (1947) reaches 60 m in height; most frequent synonym: *F.magnolioides* Borzi. It is a very popular tree for large parks and gardens, and tends to develop the characters of a banyan (viz Australian Banyan). Our cultivated specimens hardly ever exceed 15 m in height. They have a short and heavy trunk (often bent) and a wide spreading root system. Bark pale grey, rather smooth, with lenticels and shallow horizontal grooves. The heavy branches form a very wide spreading crown. Aerial roots may appear in older specimens.

Leaves long-stalked; blade entire, roughly ovate-elliptical, coriaceous, 12 to 20 cm long and 5 to 8 cm wide, dark green and shiny above and somewhat rusty beneath. Fine lateral nerve system. Terminal bud prominent, up to 10 cm long. Figs solitary or in pairs, stalked, globose, up to 2 cm in diameter, purple-coloured with pale dots. – The Moreton Bay Fig is propagated from cuttings rooted on the tree and grows best at low altitudes. It is quite resistant to saltspray and strong wind. According to Barrett (47) the wood is 'handsome but hard to season', the latex is used for rubber, and leaves are consumed by cattle.

Ref.: Barrett 47, 56; Bircher; Chittenden; Condit; Encke; Graf; Harrison; Kunkel 69; Lombardo 58; Menninger 64; Neal.

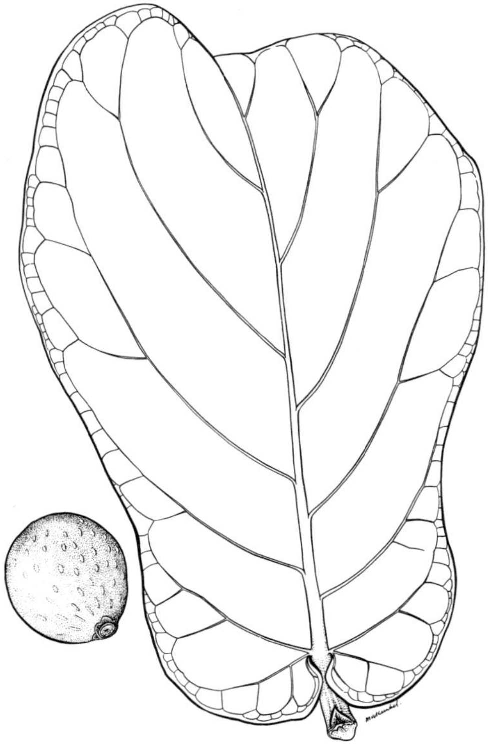

Fig. 22. *Ficus lyrata*; leaf and fruit = 1/3

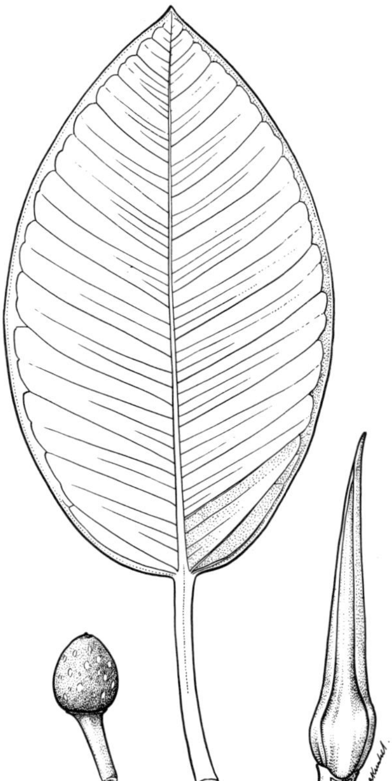

Fig. 23. *Ficus macrophylla*; details = 2/5

Ficus microcarpa Linn. fil. – Laurel Fig

A widely cultivated tree of large parks and avenues, often cited as *F.retusa* L., or *F.nitida* Thunb.; alternative common names: Indian Laurel, Malayan Banyan, and Chinese Banyan. The species is native in tropical South Asia and is now grown in most subtropical countries.

Large tree up to 20 m in height, with a dark, very wide spreading crown. Trunk relatively short, up to 1 m in diameter or even more, often irregular or ridged. Bark grey which becomes rather scaly in older specimens; branches somewhat spotty. Evergreen. Leaves short-stalked, dark green and shiny, thick-herbaceous or even somewhat fleshy, ovate-lanceolate, 6 to 8 cm long and usually up to 3 cm wide. Terminal bud short. Figs small (5–8 mm ∅), sessile, true 'fig'-shaped, yellowish-green.

Propagated from cut branches which root easily. The Laurel Fig is not recommended for small gardens as its root system can be rather destructive: 'man-made structures are hardly a match for powerful, vigorous trees. Walls crumble, sidewalks buckle, paving splits, and pipes clog and crack under pressures from growing root system' (a quotation cited by Condit). However the species is an excellent shade tree, provides a home for hundreds of noisy birds, and has its medicinal virtues too. Its latex is rich in resin; 'roots and leaves are used for applications to wounds and bruises; bark and leaves for head-ache ... and juice of the bark internally for liver disease' (Burkill). – Unfortunately specimens grown in the Canary Islands are frequently attacked by Thrips which cause deformed, curled leaves; only a heavy pruning has been found an effective control for this pest.

Ref.: (Barrett 56); (Bircher); (Burkill); Condit; (Corner); (El Hadidi & Boulos); (Encke); (Graf); Kunkel 69; (Little & Wadsworth); (Lombardo 58); Moeller 71; (Neal); (Schaeffer); Synge; (Uphof).

Fig. 24. *Ficus microcarpa*; branch and details = 1/2

Ficus cf. *natalensis* Hochst. ex Krauss.

This evidently African species, although rare in the Canary Islands, is included here because of the scarceness of illustrations available. It remains problematic, especially as our material was originally named as *Ficus thonningii* Blume; however it differs from that species by its stalked globose fruits approximately 1 cm in diameter. It seems to be related to *F.thonningii* as well as to *F.dekdekana* (Miq.) A.Rich., both of which are of West African origin.

A low-branching small tree easily recognized by its stilt roots. Trunk short, bark dark grey and somewhat rough. Deciduous. Leaves clustered in terminal position, short-stalked; blade oblanceolate, subcoriaceous, very fine-veined on the upper surface, 6 to 10 cm long and 2,5 to 4 cm wide. Terminal bud short.

Ref.: Barrett 48; Condit; Palmer & Pitman; Purseglove.

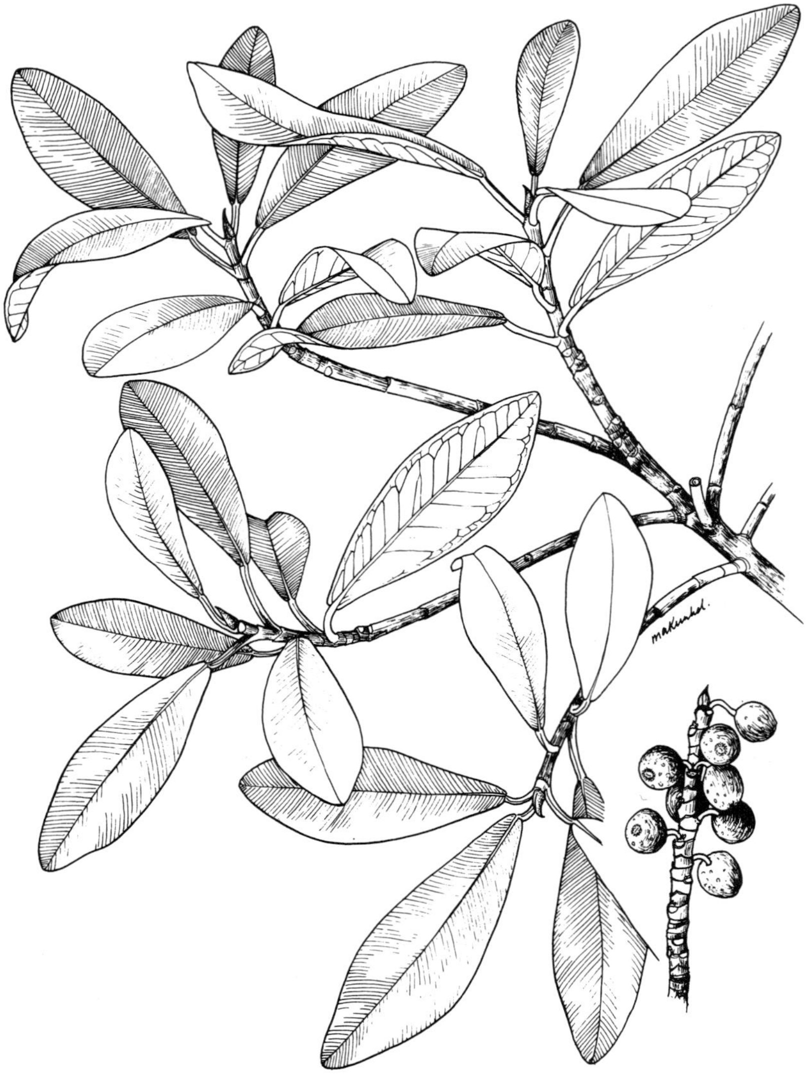

Fig. 25. *Ficus cf. natalensis*; drawings = 2/5

Ficus religiosa Linné – Bo Tree

This sacred tree native in India is recorded as one of the longest living woody species known: According to Barrett, Condit and others a rooted cutting from the holiest specimen known in India was sent in 288 B.C. to Ceylon (now Sri Lanka) where it soon grew into a proper tree, and its history has been recorded for over 2.200 years; secondary stems of this particular specimen are said to be still alive. One may feel something of the magic of the Bo Tree and its history when admiring one of the long-fallen leaves which consists of nothing else but an incredibly fine nerve system from which all leaf tissue has rotted away.
A tree up to 30 m tall with a wide spreading crown; trunk short, somewhat irregular or knotty; bark brownish-grey with lenticels and horizontal grooves. Twigs very slender, pendent; foliage evergreen or very briefly deciduous. Leaves long-stalked, stalks 6 to 12 cm long. Blade up to 10 cm wide, heart-shaped to almost triangular, ending abruptly in a very long, narrow apex. The leaves are dark green or somewhat glaucous, subcoriaceous, have a prominent nerve system, and rattle in the wind just as some poplar leaves do. Figs dark, 6 to 8 mm in diameter, usually in pairs, sessile in leaf-axils.

The Bo or Peepul Tree is best propagated from woody rootshoots and thrives in somewhat cooler, moist climates of tropical zones. There are only few specimens in these islands. It provides food for silkworms, and the wood is used for boxes, charcoal and fuel. 'Buddhists and usually Hindus refrain from cutting this species, but Mohammedans have no such inhibition' (Barrett 51).

Ref.: Barrett 51, 56; Bircher; Burkill; Chittenden; Condit; Corner; El Hadidi & Boulos; Encke; Graf; Kunkel 69; Menninger 67; Neal; Purseglove; Uphof.

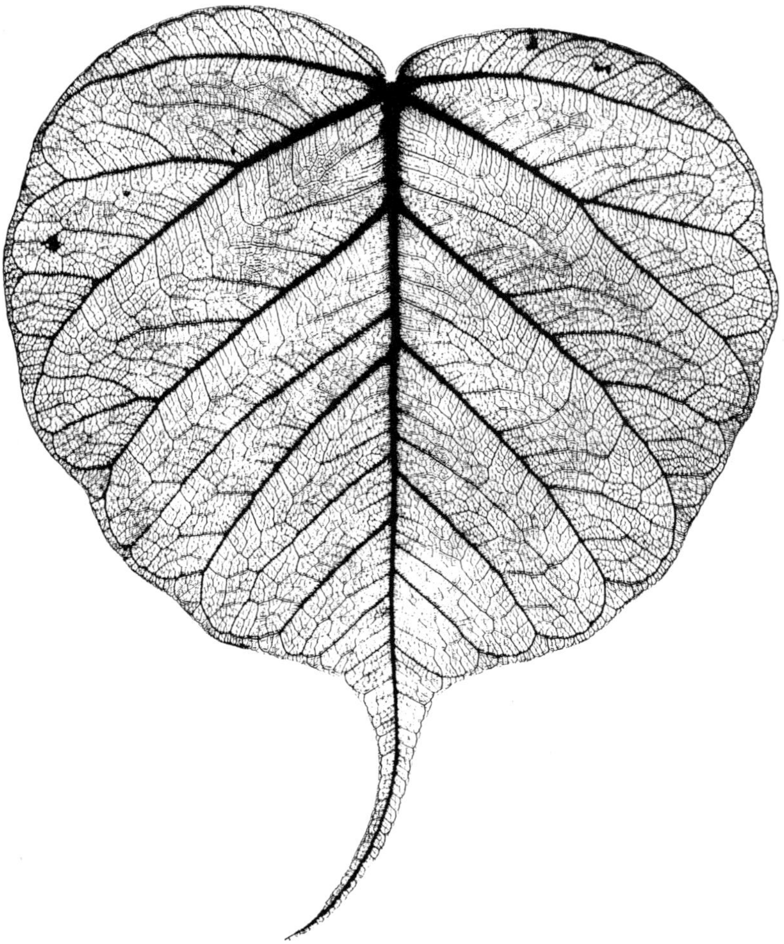

Fig. 26. *Ficus religiosa*; leaf print (remaining nerve system) = natural size

Ficus rubiginosa Desf. ex Vent. – Rusty Fig

This tree is a native of Australia and is also called Port Jackson Fig, Illawarra Fig, and Botany Bay Fig. A frequently cited synonym is *Ficus australis* Willd. – The Rusty Fig is often cultivated even in quite small gardens and seems to lack the usually feared destructive root-system. It grows well in seaside gardens, and is also planted as a roadside tree.

The Rusty Fig is reported to be a large tree of banyan type (Barrett 47); however our specimens are smaller (8 to 12 m), have a rather rounded, dense dark crown, a slender trunk and – for easy recognition – masses of thin reddish aerial roots hanging from its branches. Bark grey and rough, scaly in older specimens. Leaves stalked; blade oval-shaped, leathery, 6 to 10 cm long and 4 to 6 cm wide, dark green above and rusty-coloured beneath. Nerve system from above hardly visible. Terminal bud stout. Petiole and under surface of leaves faintly tomentose. Figs sessile, in pairs, about 1 cm in diameter, brownish, partly covered by semi-persistant bracts. – Propagation from cuttings and prepared layerings. The species is sometimes grown as a pot plant; a variegated form is also known.

Ref.: Barrett 47; Chittenden; Condit; Encke; Graf; Harrison; Kunkel 69; Menninger 64; Neal.

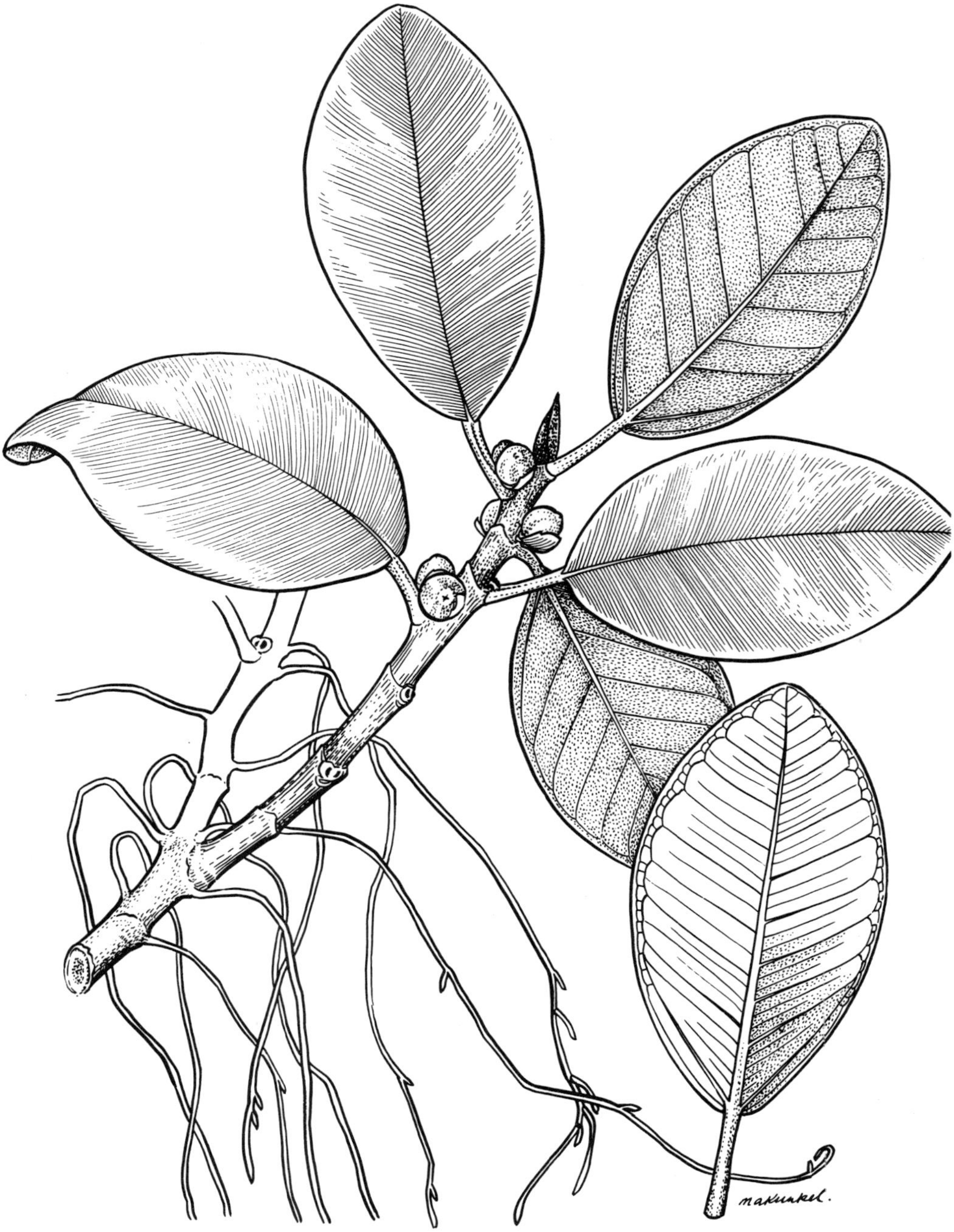

Fig. 27. *Ficus rubiginosa*; branch with aerial roots = 3/5

Ficus tuerckheimii Standley – Tuerckheim's Fig

Although this species is quite rare in our gardens, I prefer to include it in this account as very few illustrations of this tree are known to exist. Our material was kindly named by Dr. DeWolf. The common name, of course, is invented, but I was unable to find any other or more suitable name for it.

It seems to be a Central American species which is described as follows: Tree 10 to 12 m tall, with an open, spreading crown. Trunk short, 50 to 60 cm in diameter, with short buttresses; bark grey, fissured; branches somewhat knotty. Probably evergreen. Leaves long-stalked (petiole 8 to 12 cm), dark green, leathery, with prominent lateral veins; blade ovate-elliptical, with a cordate base, 13 to 17 cm long and 10 to 12 cm wide. Terminal bud 2 to 3 cm long. Figs axillary, in pairs, subglobose or somewhat pear-shaped, about 1,5 cm in diameter, and partly covered by persistant bracts.

Propagation presumably from rooted cuttings. The tree seems to grow best in a cooler and moist climate.

Ref.: Condit.

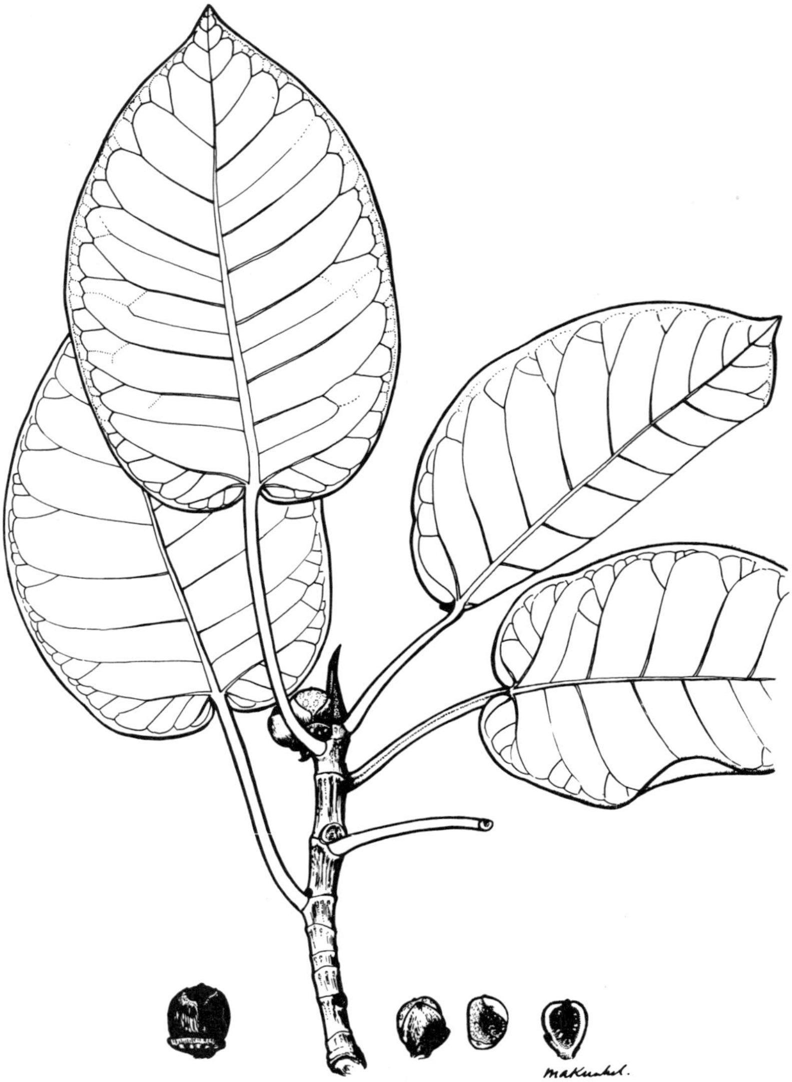

Fig. 28. *Ficus tuerckheimii*; branch and details = 2/5

Ficus urbaniana Warburg

There exists a slight mystery concerning the identity of the specimens grown in these islands. The material was, originally, introduced by Sr. M. Torres, from Venezuela, and Dr. DeWolf who kindly revised sterile material sent by the present author, tentatively confirmed the name as given above. *Ficus urbaniana* is placed as a synonym of *F.obtusifolia* H.B.K., by Little & al., whereas Condit deals with both as different species. The name *F.obtusifolia* Roxb. as cited by Uphof is a nomen nudum (see Condit) and refers to an Asiatic species.

Our specimens – to whichever species they may belong – grew into medium-sized trees with a spreading, open crown. Stem slender; bark grey and rather smooth. Young branches greenish and shiny. Foliage evergreen. Leaves long-stalked, obovate, dark green, coriaceous, up to 40 cm long and 15 cm wide, with a pale, very prominent nerve-system. Fruits not observed. – The species is propagated from rooted cuttings and grows best in a somewhat cooler climate.

Ref.: Condit; (Little & al.);
(non Uphof).

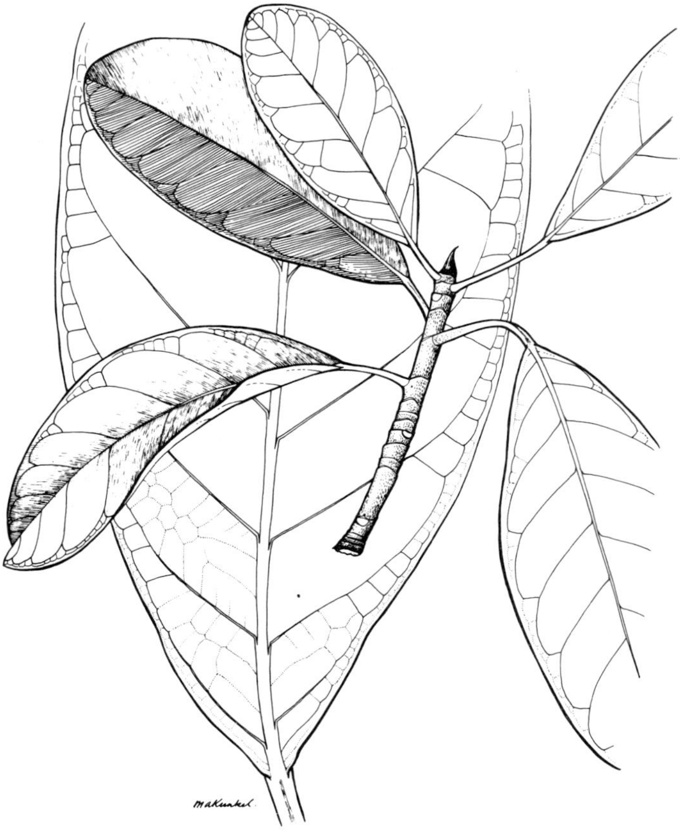

Fig. 29. *Ficus urbaniana*; leaves only (= 1/2)

Ficus virens Aiton – Spotted Fig

A rather troublesome case especially as far as its correct name is concerned: M. F. Barrett (1947) strongly believed *Ficus lacor* to be an Indian tree 'infested' by the erroneous synonym *F.infectoria* Roxb. (non Willd.), an opinion shared by M. C. Neal (1965). It remained *F.infectoria* by Bircher, Burkill, and Chittenden whereas I. J. Condit (1969), and later Corner & Stearn (in Synge), included both names as synonyms under *F.virens* Aiton. – In case anyone wants to sort out the complex problem he is invited to consult the work by Condit who dedicated over four pages to this species which I believe to be our roadside tree and described as follows:
Deciduous tree 8 to 15 meters tall; trunk short and slender, bark grey and rough. Branches spreading, terminal branchlets drooping. Terminal bud short. Leaves alternate and rather spaced, long-stalked, oval-elliptical, dark green, little leathery, ending in a blunt, mucronate apex; blade 8 to 15 cm long and 4 to 6 cm wide. Figs almost sessile, rounded, 1–1,3 cm in diameter, spotty, rather crowded in subterminal position.
This species makes a good street tree of towns at lower altitudes and is also planted in smaller gardens. According to Barrett (56) it is often seen as a shade tree in coffee plantations; the wood is used for timber, the fibre of the bark is made into ropes, and leaves are eaten by cattle.

Ref.: (Barrett 47, 56); (Bircher); (Burkill); (Chittenden); Condit; (El Hadidi & Boulos); (Neal); Synge; (Uphof).

Very closely related to *Ficus* and often united with this genus is *Coussapoa dealbata*, a large banyan tree from tropical South America of which an impressive specimen can be admired in the Orotava Botanical Garden, Tenerife. – Strangely enough the genus *Coussapoa* is placed in the Urticaceae, by Airy Shaw in Willis.

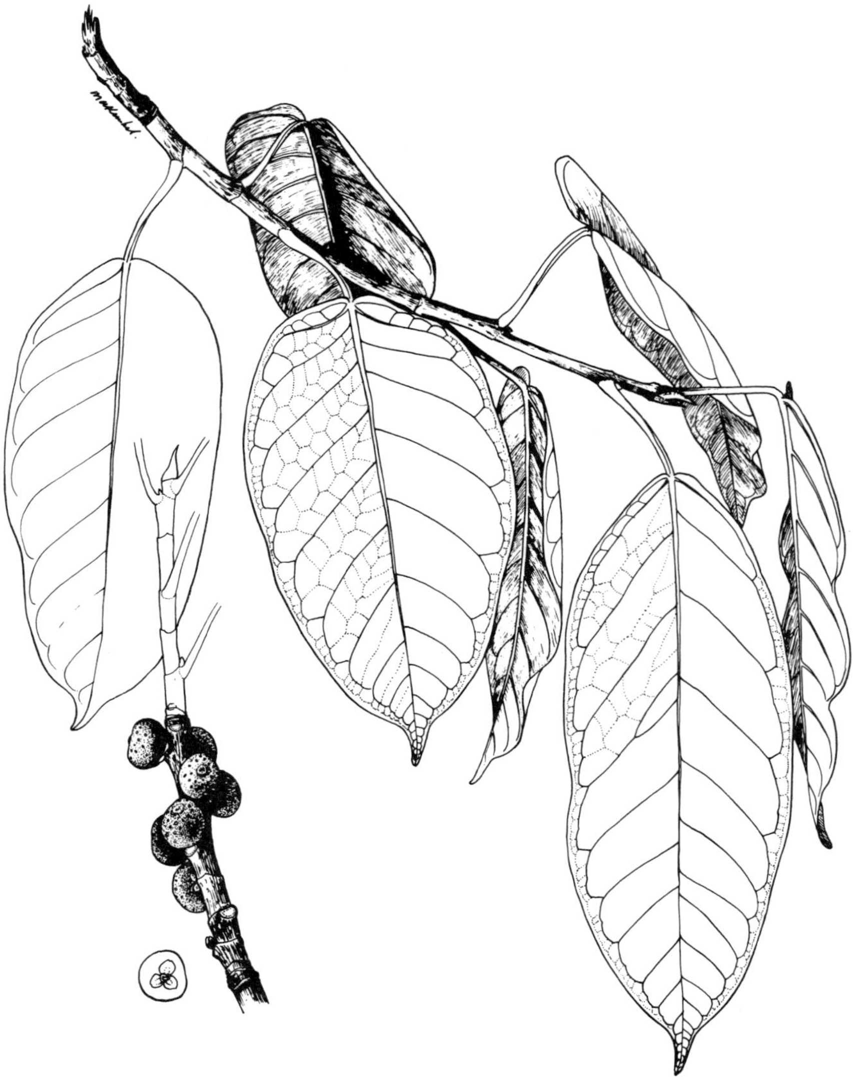

Fig. 30. *Ficus virens*; drawings = 2/5

Maclura pomifera (Raf.) Schneider – Osage Orange

A North American tree also known as 'Bow wood'; locally naturalized in Southern Europe (Tutin 64). The species can be grown as an effective hedge plant but may also grow into a medium-sized tree. A very vigorous plant, with respectable axillary spines, and in some places feared as a 'weed'.

Tree up to 15 m tall. Trunk usually branched from near to the base; branches upright, armed with spines 2 to 4 cm long. The species is deciduous. Leaves alternate, almost opposite, or in whorls; petiole 2 to 4 cm long. Blade ovate to oblong-lanceolate, herbaceous, bright green, pointed at apex; 6 to 15 cm long. Flowers small and greenish; male flowers in long-stalked, rounded clusters; female ones developing into globose fleshy fruits (syncarps) of a golden-yellow colour, up to 10 (15) cm in diameter.

The fruit evidentally is not edible but 'contains 10% antioxidative compounds' (Uphof) and is used in cosmetics, lubrication-grease etc. The wood is very strong and durable and was used by American Indians for making bows; bark of roots yields a yellow dye. The species is propagated from seeds, layers and cuttings, and grows best in cooler climates. A spineless form is mentioned by Chittenden.

Ref.: Bean; Bircher; Chanes; Chittenden; Encke; Fitschen; Lombardo 58; Mitchell; Tutin 64; (Uphof).

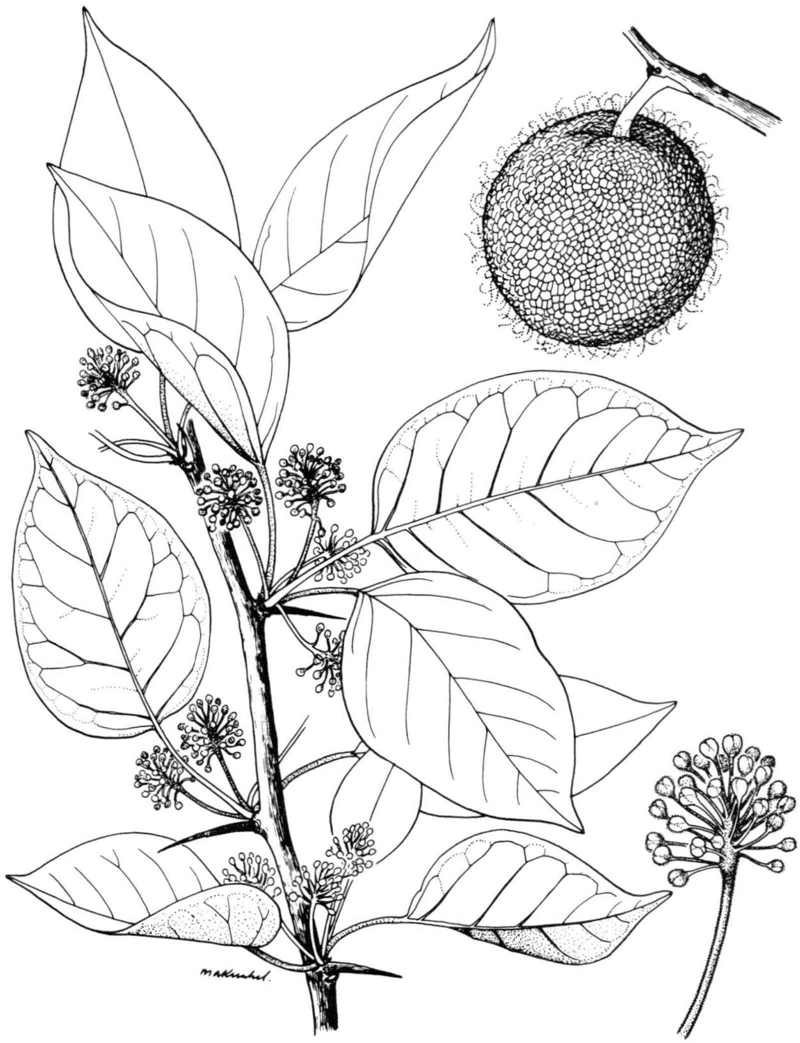

Fig. 31. *Maclura pomifera*; flowering branch and fruit = 2/5, solitary inflorescence = 4/5

Morus alba Linné – White Mulberry

This frequently planted species is a native of China, and is now naturalized in many countries, even of Southern Europe. It is a small tree (up to 10 m tall), with a short trunk; bark greyish-brown and rather rough. Crown quite open. The species is deciduous. Leaves long-stalked, herbaceous, glabrous, about heart-shaped, up to 20 cm long and 15 cm wide, with a dentate or lobed margin (varieties!). Latex present as in fig trees. Flowers unisexual, inconspicuous, greenish, in short cylindrical spikes. The 'false fruits' (syncarps, or fruit clusters) are fleshy, whitish or reddish, and are edible although rather tasteless.

A quite variable species of which a number of forms and cultivars are grown in gardens including a form with pendulous branches (see Bean, Chittenden, Fitschen and others). The White Mulberry has been cultivated 'since time immemorial as a food plant of the silk worm' (Uphof). Propagation from cuttings and layers; best grown in cooler zones. Roots and leaves are diuretic, and bark and roots are said to be vermifuge (Font Quer). The wood is hard and workable.

Ref.: Bean; Bircher; Chanes; Chittenden; El Hadidi & Boulos; Encke; Fitschen; Font Quer; Kennard & Winters; Kunkel 69; Lombardo 58; Menninger 64; Mitchell; Neal; Polunin; Polunin & Smythies; Ruiz de la Torre; Tutin 64; Uphof.

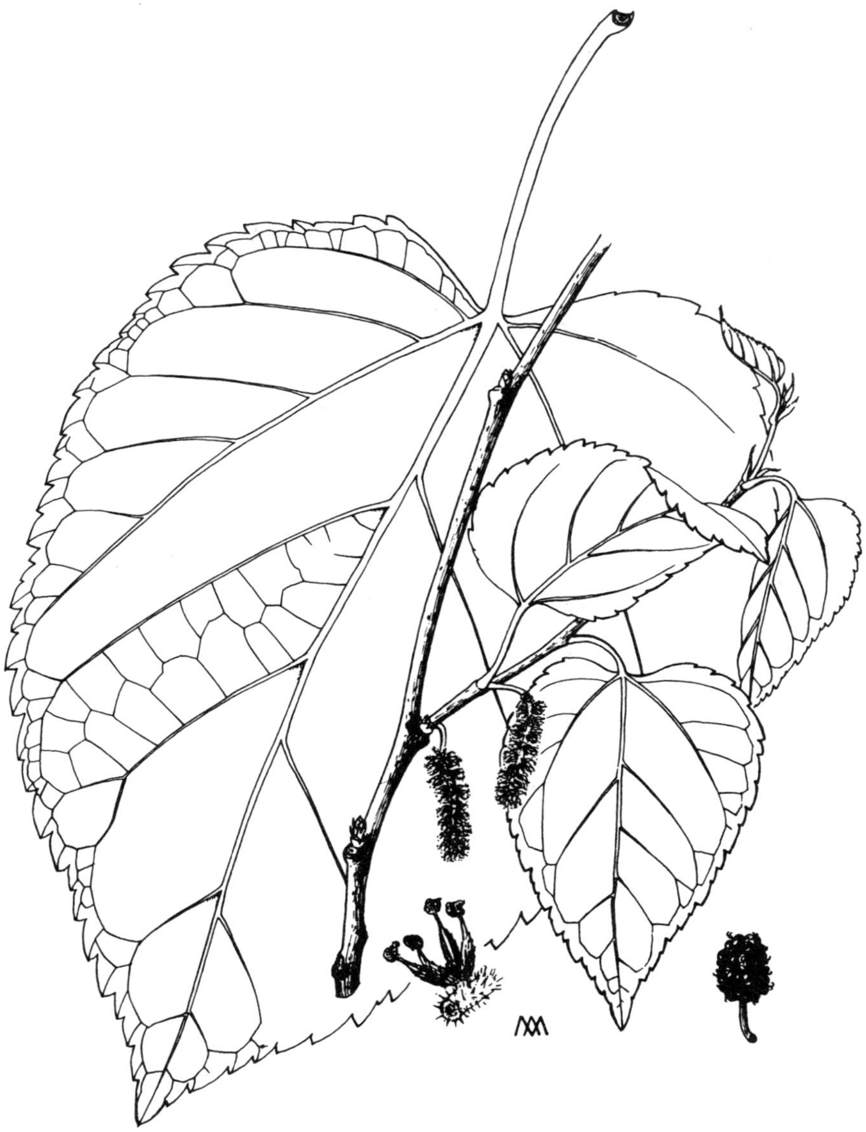

Fig. 32. *Morus alba*; leaf and details = 2/5 from Kunkel: 'Arboles exóticos'

Morus nigra Linné – Black Mulberry

Also of Asiatic origin (C.Asia), this somewhat smaller but more robust, only briefly deciduous species is mainly cultivated as a shade tree and for its 'fruits' (syncarps) which are eaten raw or made into jam and syrups. According to Uphof the 'fruits are also source of a rouge, used in cosmetics'. The Black Mulberry has been cultivated in Mediterranean countries for some thousands of years and is considered to be well naturalized in many places including these islands.

Tree 6 to 10 m in height; trunk short, up to 80 cm in diameter. Bark grey and rough, very scaly in older specimens. Crown widespreading. Leaves petiolate, ovate to heart-shaped, coarsely toothed, subcoriaceous, dark green, with short hairs on the upper surface, and somewhat pubescent beneath; blade up to 20 cm long and 16 cm wide. Spikes cylindrical, of which the shorter female inflorescences produce the fleshy, edible, dark purple or almost black syncarps.

Propagation from cuttings but naturally also from seeds. The species thrives best at lower altitudes and is often seen growing in fields together with the Common Fig. A preparation of the bark makes a laxative and is said to serve for intestinal worms.

Ref.: Bircher; Chittenden; Encke; Fitschen; Font Quer; Kunkel 69; Little & al.; Long & Lakela; Mitchell; Polunin & Smythies; Ruiz de la Torre; Tutin 64; Uphof.

Other species and also very famous (but in these islands only in the Botanical Garden of Tenerife) include the Bread-fruit tree (*Artocarpus altilis* [Park.] Fosberg), and *A.heterophyllus* Lam., the Jak-fruit tree which are dealt with in more specialized 'floras'; for their geographical distribution, economic importance and other data please consult Cobley, Molesworth Allen, Purseglove and other authorities.

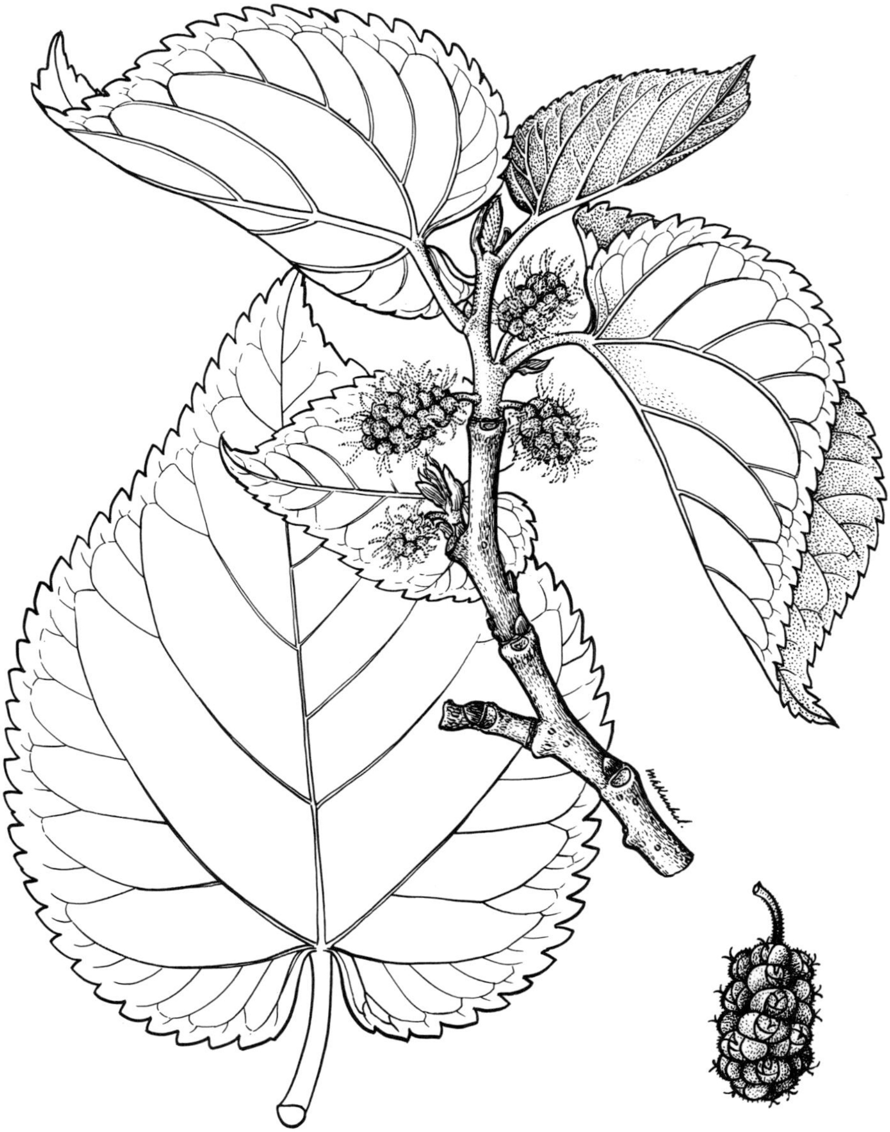

Fig. 33. *Morus nigra*; branch and details = 4/5

CASUARINACEAE, the Casuarina family

This curious family of trees and shrubs with its inconspicuous flowers is native in Southeast Asia, parts of Australia, and some Polynesian islands. There are only 2 genera, with 65 species, of which 45 belong to the genus *Casuarina*. These numbers are given in the new edition of Willis, whereas most other authors (viz Takhtajan, Rendle, and common Compendia) mention the family as monogeneric, with 25 to 40 species only. – Several species are widely planted for reafforestation of semi-arid zones.

Casuarina equisetifolia Linné – She-oak, or Horsetail Casuarina

This tree hardly ever exceeds 20 m in height in these islands; however Neal reports 'a height of 120 to 150 feet, the remarkable growth-rate of one being 80 feet in 10 years'. Trunk erect, slender, with a greyish or greyish-brown and scaly bark. Crown pyramidal; foliage evergreen but almost continually shedding parts of its branches. Twigs more or less whorled, needle-like, jointed as in *Equisetum* and easily breaking off. Leaves reduced to scales, found as a toothed collar at the joints of branchlets which have taken over the functions of leaves and which shed gradually just as leaves normally would. Male flowers minute, in terminal cylindrical inflorescences; female flowers in short-stalked heads which then become transformed into hard cones up to 1,2 cm in diameter. Seeds small and somewhat winged; they are very light, and Little & Wadsworth report up to 300.000 seeds per pound. The wood is known as Common Ironwood which may refer to the dark-brown heartwood. It is used for fenceposts and beams, charcoal and fuel. The bark is used for tanning and as a dye, and medicinally for diarrhoea and

* According to a recently received publication by F. R. Fosberg & M.-H. Sachet (Smithsonian Contributions to Botany 24; 1975) the correct name for the well-known She-Oak seems to be *Casuarina litorea* (Rumph.) L. against *C.equisetifolia* L. as its most common synonym. However as the type-specimen of Rumphius was originally cited as 'Casaarina litorea', the present author prefers to await the results of an investigation announced by Fosberg & Sachet (1.c., p. 4). Furthermore, Fosberg, Falanruw & Sachet, in another paper also published in 1975, cited *Casuarina equisetifolia* L., without any comment, as a valid name.

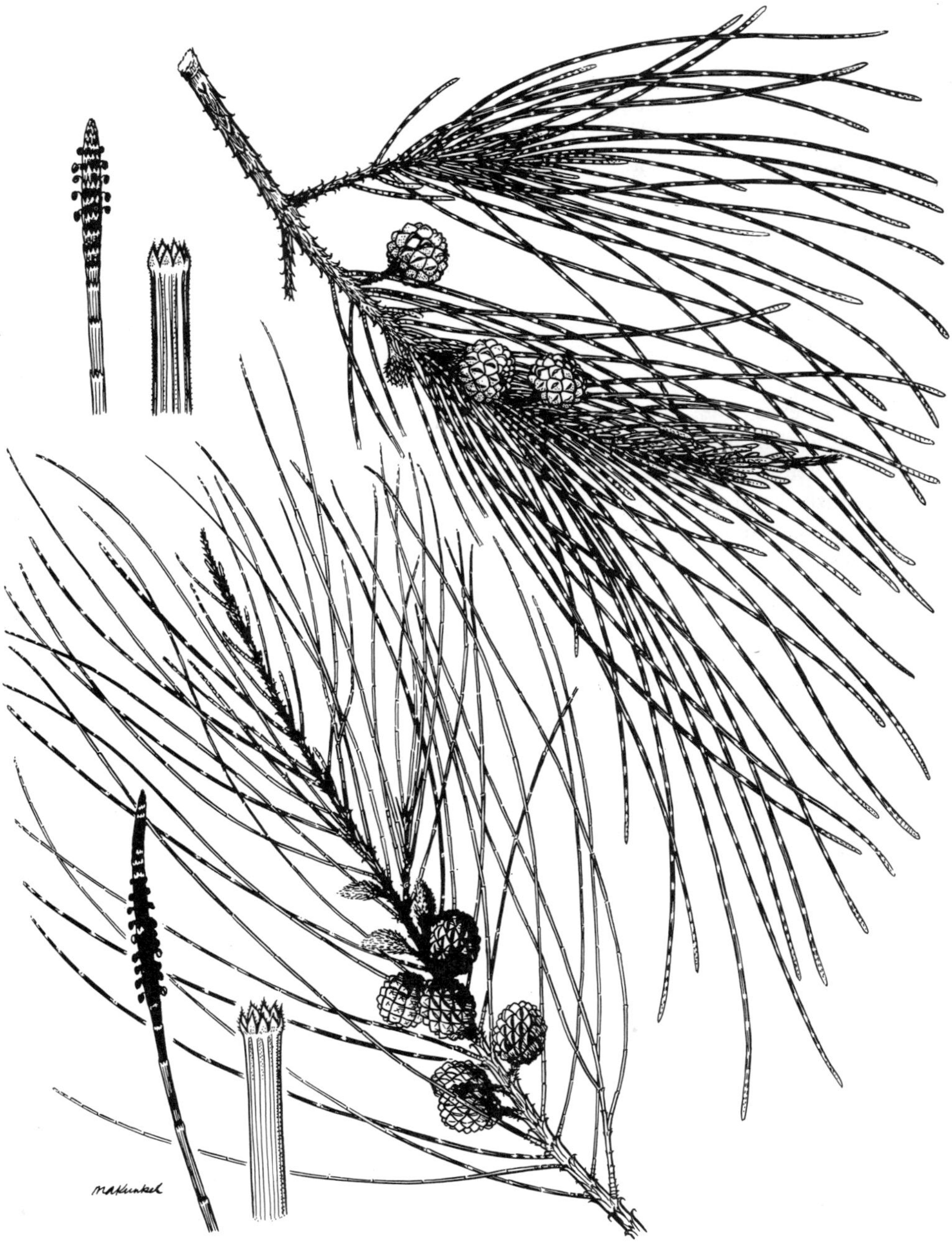

Fig. 34. *Casuarina equisetifolia* (above) and *C.cunninghamiana* (below); fruiting branches = 3/5, male inflorescence = natural size, leaf detail = × 4

dysentery (Burkill). The tree makes good hedges and an effective windbreak but hardly tolerates other vegetation nearby. It grows best near to the sea or at low altitudes. Neal mentions several mysterious powers attributed to the Casuarina. Irvine calls it the 'Whistling Pine', or 'Beefwood', and the name 'Willow' is given additionally by Adams, whereas Bircher adds 'Mile tree', and 'South Sea Ironwood'. Another common name is 'Australian Pine'.

Ref.: Adams; Barrett 56; Bircher; Burger; Burkill; Chittenden; Degener; Irvine; Keay & al.; Kunkel 69; Little & Wadsworth; Long & Lakela; Menninger 64; Moeller 71; Neal; Pesman; Schaeffer; Uphof.

Casuarina cunninghamiana Miq. – Australian River Oak

This species is of lesser importance in cultivation although I would call it more graceful than the Common She Oak. It grows faster and reaches up to 20 m in height, the trunk is heavier, and it provides more effective shade. It is frequently planted along roadsides. Branches upright and then spreading, twigs arched, longer and darker than for the one previously described. Male inflorescence up to 3 cm long, the female more oblong and more densely grouped.

All Casuarinas are easily grown from seeds. Some might also strike from rootstocks or cuttings. Our second species thrives further inland up to almost 1.000 m above sea level. It makes a graceful solitary tree in parks and gardens.

Of more recent introduction are *Casuarina stricta* Ait. (Shingle Oak), and *C.glauca* Sieb., both still little known in these islands.

Ref.: Barrett 56; Bircher; Burger; Chittenden; Corner; Harrison; Kunkel 69; Little & al.; Lombardo 58; Menninger 64; Ruiz de la Torre; Uphof.

FAGACEAE, the Beech family

According to 'Willis Dictionary' 8 genera with some 900 species of trees and shrubs of temperate and subtropical climates. Many valuable timber trees of which the boreal genus *Fagus* and the austral *Nothofagus* have never proved successful in our gardens.

Castanea sativa Miller – Spanish Chestnut

Much cultivated tree native in the more humid parts of the Mediterranean region. In the Canary Islands now partly replacing the native laurel forest between 500 and about 1.400 m above sealevel. Also planted along roadsides. Deciduous tree up to 35 m tall but usually smaller (8–15 m). Trunk short and heavy, sometimes with basal shoots; bark brownish or dark grey, furrowed or scaly; old trunks often twisted as shown by Menninger (1967: 180). Crown spreading and dense in cultivated specimens. Leaves alternate, stalked, herbaceous or subcoriaceous, oblong-lanceolate, coarsely dentate, up to 20 (25) cm long and 3 to 5 cm wide. Flowers fragrant, small, pale green or creamy; male flowers in clusters of narrow spikes up to 15 cm long; female ones solitary or grouped, very short stalked. Fruit globose and very prickly, with one large or 2 to 4 smaller roundish, shiny brown seeds.

Chestnuts are a good source of food especially when roasted or boiled; they also make a substitute for coffee, can be made into flour, added to soups and bread, used in different desserts ('Marron glacé'), and are also a source of oil (Uphof). The wood makes a good fuel and is locally used in carpentry, made into railway sleepers and fence posts. The bark is used for tanning. The Spanish Chestnut makes a fine solitary specimen in parks, and most remarkable samples are cited for Britain by Bean: 118 feet tall, maximum girth of trunk over 25 ft. Several varieties are known to be in cultivation elsewhere. Propagation from seeds, cuttings and layers. Referring to the cultivation of this species see also McKay & Jaynes. – In some European gardens also *Castanea crenata* Sieb. & Zucc., the Japanese Chestnut, which is smaller than the species described above.

Ref.: Adams; Bean; Chanes; Chittenden; Dimitri & Milano 50; Encke; Fitschen; Font Quer; Harrison; Lombardo 58; McKay & Jaynes; Menninger 67; Mitchell; Ruiz de la Torre; Tutin 64; Uphof.

Fig. 35. *Castanea sativa*; drawing and details = 1/2

Quercus spp. – The Oaks

There is no native species of oak in the Canary Islands, although one (from the Western Mediterranean region) bears the misleading name *Quercus canariensis* Willd. The genus consists of about 450 species most of which are native in countries of the Northern Hemisphere. Most oaks are hard-wooded and valuable timber trees. In these islands only 3 species frequently planted

Quercus ilex Linné – Holly Oak

Evergreen Mediterranean tree up to 25 m in height. Trunk usually short and soon branching; branches heavy, upright or spreading; bark dark and fissured. Leaves alternate, petiolate, oblong-ovate or obovate, coriaceous, glossy dark green, dentate-serrate or, sometimes, almost entire; 3 to 5 cm long and 2 to 3 cm wide. Flowers yellowish green and very small; male flowers in narrow spikes (catkins) 3 to 5 cm long, female flowers usually solitary in axillary position. Fruits oblong-cylindrical, up to 3 cm long, brownish glossy, partly covered by a fine-scaled cupule.

Wood very strong and heavy, used for furniture. Acorns (nuts) are edible. A specific leaf-gall seems to be the source of tanning material. The species is recommended for seaside plantations by Menninger.

Ref.: Chanes; Chittenden; Dimitri & Milano 50; Encke; Fitschen; Font Quer; Harrison; Lombardo 58; Menninger 64; Mitchell; Ruiz de la Torre; Schwarz; Uphof.

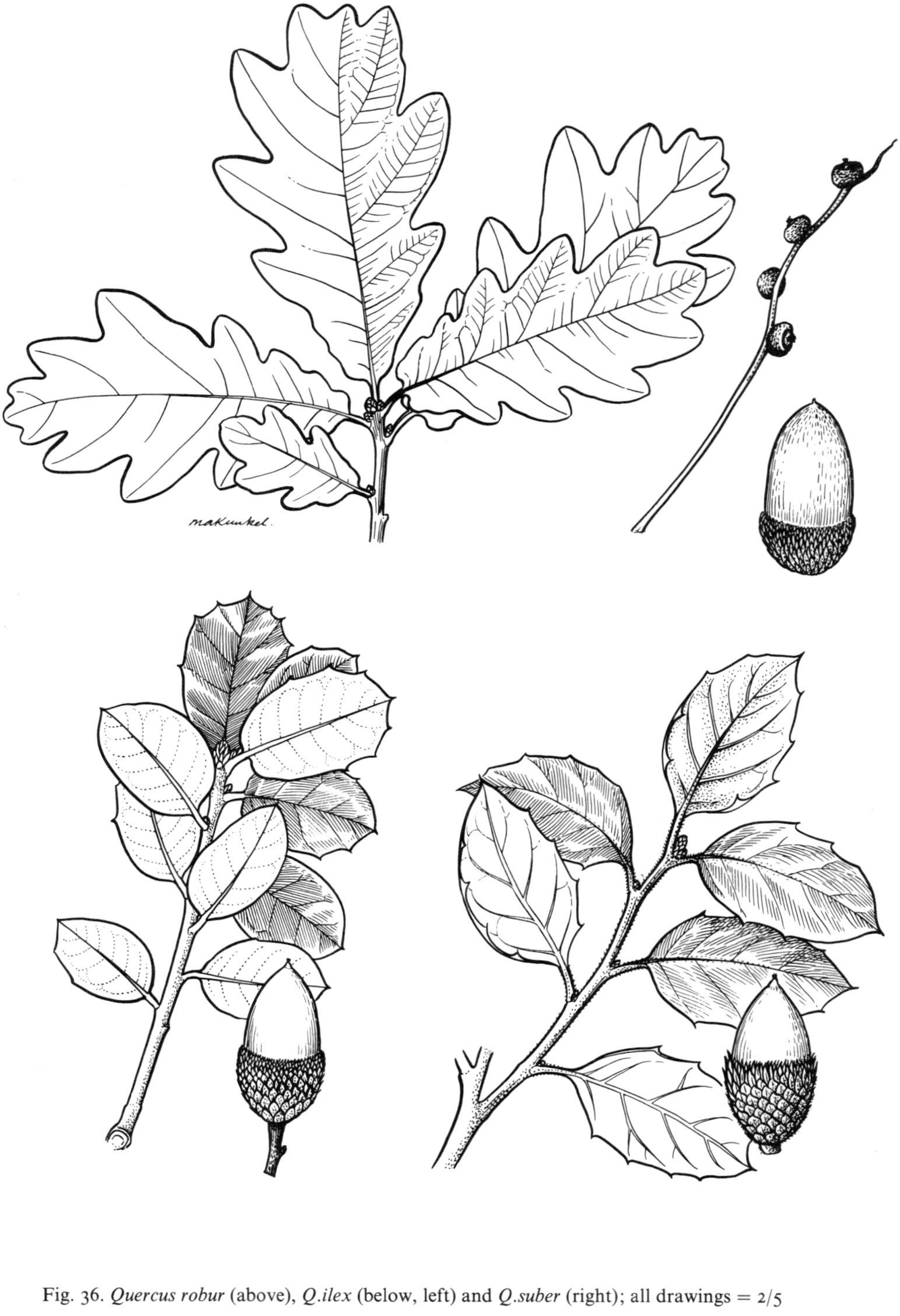

Fig. 36. *Quercus robur* (above), *Q.ilex* (below, left) and *Q.suber* (right); all drawings = 2/5

Quercus suber Linné – Cork Oak

Native and much cultivated in the Central and Western Mediterranean region, and the source of the cork of commerce. An evergreen tree up to 20 m in height, with a pronounced stem which is usually straight and easily recognized because of the thick bark which is cut or pealed off. Twigs tomentose. Leaves coriaceous and rather polymorphic, usually oblong, 4 to 7 cm long and up to 3 cm wide; entire or distantly dentate-spiny. Male flowers spaced on spikes up to 10 cm long, female flowers short-stalked and solitary or clustered. Fruits as in the above described species but the cupule shows a fringed border.
Both species cited above belong to drier countrysides, whereas the following is proper to higher and more humid zones. – According to Huxley (p. 76) 'the outer bark of the Cork Oak can be up to 20 cm thick'.

Ref.: Chanes; Chittenden; Dimitri & Milano 50; Encke; Fitschen; Font Quer; Harrison; Huxley; Lombardo 58; Mitchell; Moeller 68; Neal; Rauh; Ruiz de la Torre; Schwarz; Uphof.

Quercus robur Linné – Common Oak

This is probably the most common species cultivated in these islands. It is native in Eurasia and, according to O. Schwarz, grows in 'most of Europe, except the extreme north and parts of the Mediterranean region'. The Common Oak propagates freely from seeds and suckers.
Large deciduous tree up to 40 m tall; trunk stout but often irregular in shape, reaching 2 m in diameter or more. Bark greyish and somewhat fissured at first, later brownish and deeply furrowed. Crown large and spreading, branches usually covered by mosses and lichens. Leaves alternate, stalked, herbaceous, glabrous, up to 10 (12) cm long and 5 to 8 cm wide, obtusely lobed; nerves pinnate. Flowers small and greenish; male flowers in short spikes 6 to 8 cm long; female ones sessile, solitary or in small clusters. Fruits (acorns) oblong-cylindrical, solitary or several together on a common stalk; cupule hard-scaly.
The Common or English Oak is frequently used for reafforestation. The wood is yellowish-brown and heavy, easy to split and very durable, resistant to water and used for carpentry, furniture, floors, shipbuilding etc. (Uphof). It is the source of acetic and tannic acid. Acorns are used to feed pigs and are also a substitute for coffee. According to Encke and Chittenden there are many cultivars grown in European parks, and a hybrid with the Holly Oak has been described as *Quercus × turneri* Willd.

Ref.: Chanes; Chittenden; Dimitri & Milano 50; Encke; Fitschen; Font Quer; Harrison; Lombardo 58; Mitchell; Ruiz de la Torre; Schwarz; Uphof.

JUGLANDACEAE, the Walnut family

A family of 6 or 7 genera with almost 50 species, natives mainly of temperate countries in the Northern Hemisphere. Mostly trees which are deciduous and grow at higher and moister altitudes. In cultivation in these islands only two species:

Carya illinoensis (Wang.) K.Koch – Pecan

This is a small tree native in North America, and by no means common in Canary gardens. According to Elwes & Henry 'attaining in America 170 feet in height and 18 feet in girth'. Trunk straight and slender (in our specimens); bark dark grey and fissured. Branches upright forming a loose pyramidal crown; branchlets slightly pubescent. Leaves imparipinnate, up to 30 cm long, with 7 to 15 pairs of herbaceous, lanceolate leaflets, with serrate margin and an unequal base; the upper leaflet may reach up to 12 cm in length. Flowers greenish and small, in short or longer racemes according to their sex. Fruits: ovoid or oblong edible nuts.

The species thrives best in cooler climates and is propagated from seeds or root-shoots. According to Uphof nuts are eaten raw or in candies and cakes. The wood of this tree is used for fuel and for agricultural implements. For further orientation on the Pecan see also Madden & al.

Ref.: Bean; Bircher; Dimitri & Milano 51; (Elwes & Henry); Fitschen; Kunkel 69; Lombardo 58; (Chittenden); Madden & al.; Martinez; Uphof.

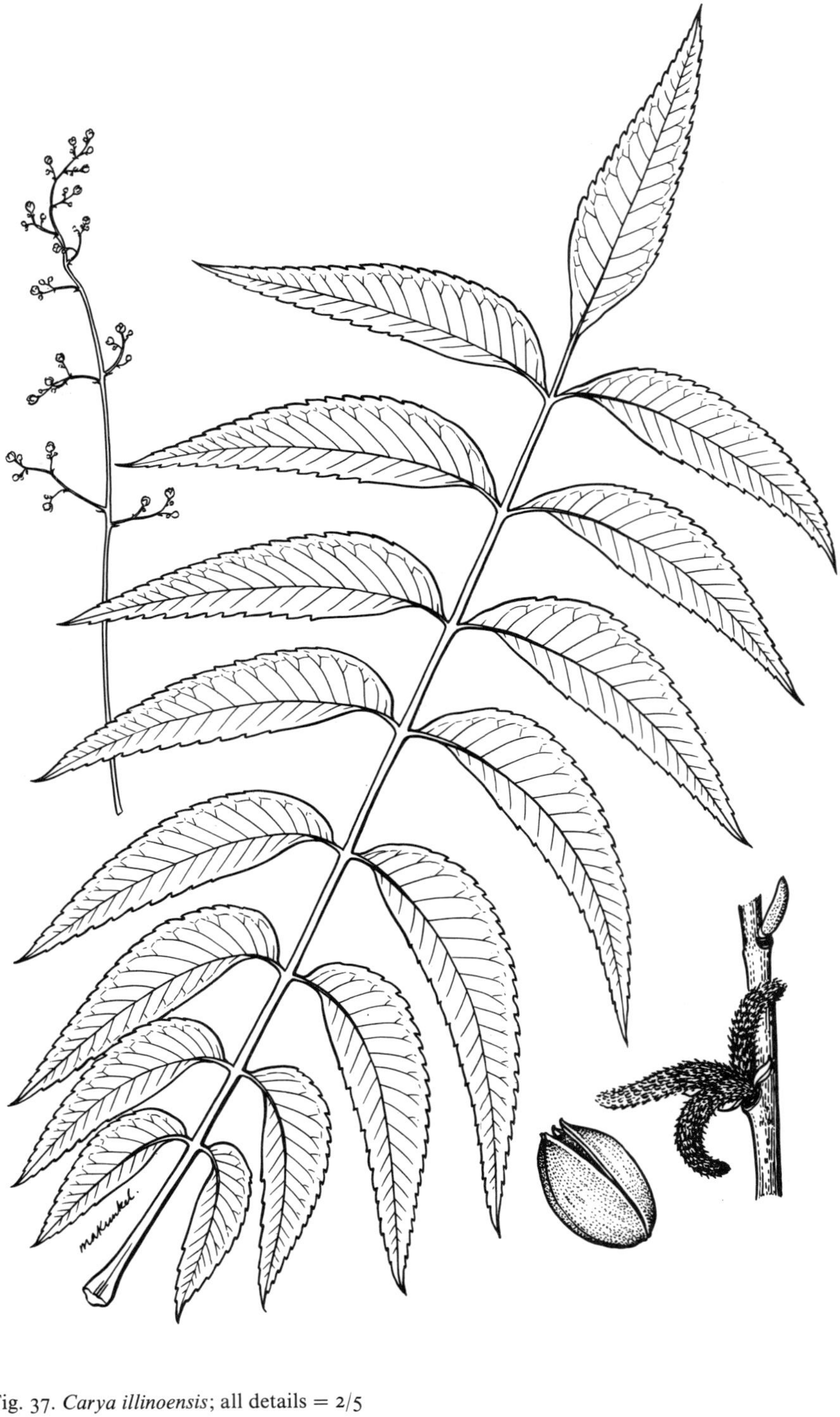

Fig. 37. *Carya illinoensis*; all details = 2/5

Juglans regia Linné – Common Walnut

Probably native at medium or higher elevations from South-eastern Europe through parts of the Middle East reaching China, and much cultivated for its fruits and the highly appreciated, workable wood. Only above 500 meters in the cloud zone of these islands where the species is locally abundant.

Tree up to 15 (20) m tall. Trunk short, in old specimens over 2 m in diameter; bark greyish, fissured or furrowed. Crown spreading; branches usually densely covered by lichens. Leaves imparipinnate, 20 to 40 cm long, with 5 to 9 leaflets which are herbaceous, ovate, acute and up to 10 cm long. Flowers small, greenish; male flowers in compact, hanging cylindrical spikes, the female flowers are solitary or grouped, and short stalked. Nuts 1 to 3, hard-shelled, more or less oval-shaped, 2,5 to 4 cm in diameter; edible and much appreciated.

According to Uphof oil obtained from the nuts is used in food processing or if dry pressed, for soap and paints. Leaves are used for dyeing. Hulls of fruits are employed in home medicine as astringent, antiscrofulous, blood-cleaning and for intestinal worms. The wood is used for furniture making. The species is propagated from seeds and rootstocks. – For varieties and cultivars see W. J. Bean. Bircher calls this species the Persian or English Walnut. As mentioned by Chittenden it seems to be involved in much hybridization.

A second species – *Juglans cinerea* L., the Butternut – is found in the Tenerife Botanical Garden only. The Butternut is native in eastern North America.

Ref.: Bean; Bircher; Chanes; Chittenden; Dimitri & Milano 51; Encke; Fitschen; Font Quer; Lombardo 58; Martinez; Mitchell; O'Rourke; Rauh; Ruiz de la Torre; Serr; Tutin 64; Uphof.

Fig. 38. *Juglans regia*; drawings = 2/5

PHYTOLACCACEAE, the Pokeberry family

A small family of about 12 to 15 species of herbs, shrubs and trees mainly from tropical America and Africa. Few species in cultivation, and the only tree of importance here is

Phytolacca dioica Linné – Ombú

Although known as Pokeberry, the South American name 'Ombú' deserves priority. Neal reports that 'in Argentina, anyone who sleeps under an ombú must beware of evil magic'. However the tree is praised in songs of Pampa singers, appreciated as a shade tree and besides other things it may provide a bed for lovers and a place for the homeless. Menninger, in his remarkable book called 'Fantastic trees' dedicates a small chapter to 'the Revered Ombú' and found it 'the perfect tree for public parks', and 'it never has to be replaced because it seemingly never dies'.
I must agree: the species is remarkable. The rootsystem of an old specimen grown in the Orotava Garden (Tenerife) provides sitting space and shelter for more people than are ever found at one and the same time in this botanical garden; I think a good hundred people could easily sit in the shade of this particular tree, using for seats only the spreading base which forms ponderous benches. However, if we come to analyze our Ombú more thoroughly, anatomists will complain that it is not a tree but a giant perennial herb: the trunk shows no annual rings but consists of a fibrous structure similar to that of palms or the Pawpaw. In spite of whatever is said, the Ombú reaches 15 to 20 m in height. Trunk short or often branched from the base which is irregular in shape and somewhat inflated. Bark grey or yellowish-grey, fissured and, later, scaly. Branches upright then arching or spreading, rather brittle. Twigs finely pubescent. Evergreen or very briefly deciduous. Leaves alternate, oblong-ovate, herbaceous, long-stalked; with prominent nerves. Leaves 6 to 10 cm long; in young specimens 3 to 5 times larger. Male flowers cream-coloured, in showy racemes up to 15 cm long; female flowers much smaller. Fruits ('poke-berries') fleshy, ribbed, flattened, 7 to 10-seeded, yellowish when ripe, in hanging racemes.
Propagation from seeds and cuttings, from the latter relatively fast growing. 'Wood' spongy and apparently of no use. Decorative tree for larger parks and gardens.

Ref.: Chanes; Chittenden; Encke; Garcia; Kunkel 69; Lombardo 64, 68; Menninger 67; Milano 64; Moeller 68; Neal.

Fig. 39. *Phytolacca dioica*; branch and details = 2/5

POLYGONACEAE, the Buckwheat family

Although this family consists of about 40 genera with some 750 to 800 species, except *Triplaris* and – up to a certain point – *Coccoloba*, all others are shrubs, herbs, or vines. The family has provided us with some remarkably persistant weeds such as *Emex*, *Polygonum* and *Rumex*, as well as the Rhubarb (*Rheum*); species of *Homalocladium*, *Muehlenbeckia* and *Antigonon* are grown as ornamentals. – In these islands about 20 species are present, mainly introduced weeds; however *Rumex lunaria* and *R.maderensis* are endemic shrubs. Referring to exotic trees we may limit the account to the

Coccoloba uvifera (L.) Linné – Sea-grape

'At home in thickets along sandy shores in warm parts of America' (Neal), this species is widely cultivated, especially in seaside gardens, and makes an admirable ornamental with its attractive leaves. According to Little & Wadsworth 'very probably seagrape was the first land plant of America seen by Christopher Columbus ...'.

It is a small, straggly or spreading tree up to 10 m in height, with a trunk 20 to 30 cm in diameter; bark greyish, thin, smooth or pealing off in flakes. Branches with membranous sheaths; easily breaking off. Evergreen, or very briefly deciduous. Leaves alternate, short-petioled, rounded to kidney-shaped, leathery, smooth-margined or somewhat undulate, up to 25 cm broad; red-veined. They are usually greyish green but a remarkable glossy-red when young. Flowers small, cream-coloured, in narrow racemes 20 to 30 cm long. Fruits subglobose or drop-shaped, red-purple when ripe, up to 1,5 cm in diameter.

Grown from seeds or cuttings, this species also makes a good natural fence and is planted along roadsides. The wood is used for posts and fuel, and is also reported to be useful for furniture and cabinet-making. When boiled it yields a red dye (Neal). Most authors say that early colonists used fresh leaves as a substitute for paper when writing their messages. 'Jelly and a wine-like beverage can be prepared from the fruits, which are also eaten raw' (Little & Wadsworth). The roots have their application to cure dysentery, and a gum from the bark is used for tanning and also for throat ailments.

Ref.: Adams; Bircher; Chittenden; Degener; Encke; Graf; Kunkel 69; Little & Wadsworth; Long & Lakela; Martinez; Menninger 64; Neal; (Uphof).

Fig. 40. *Coccoloba uvifera*; drawing = 2/5

CLUSIACEAE, the Mamey family

Leaving out the Hypericaceae s.str. excluded by Hutchinson, Takhtajan and some other authors, less than 30 genera with about 450 species now remain in this family, being mostly trees and shrubs of tropical zones. Most have showy flowers, some have edible fruits, and nearly all are characterized by opposite, rather large leathery leaves, and abundant whitish, yellow or even orange-coloured latex. African species of *Mammea*, *Symphonia*, *Allanblackia*, *Pentadesma* etc. are showy trees indeed but have hardly ever been successful in cultivation. The commonly cultivated American species of *Calophyllum* and *Clusia* are not found in Canary gardens. Here only and uncommon enough

Mammea americana Linné – Mamey

The Mammee-apple, as it is named in the Anglo-American language, is a handsome evergreen tree native in the Caribbean region but cultivated now in most tropical countries. It is a tree up to 20 m high, with a pronounced trunk 50 to 70 cm in diameter. Bark greyish, scaly in older specimens; branches with lenticels. Leaves obovate or oblong-elliptic, dark green, glossy, leathery, up to 25 cm long and 10 cm wide, with fine lateral veins and very small gland dots. Flowers fragrant, single or grouped on branches as well as appearing on the trunk (cauliflor); male and female flowers separated but they may appear on the same tree; petals white. The large fruit is judged to be a berry; it is globose, thick-skinned, fleshy, from 8 to 20 cm in diameter, and is edible. The species is propagated from seeds and cuttings and grows best in moist places of the warmer zone. The wood is used for fence posts and for fuel, the fruits are eaten raw or made into preserves. According to Neal the destillation of flowers provides a liquor: 'Eau (or Crème) de Créole' (see also Little & Wadsworth). The seeds are said to be poisonous and are used (powdered) as an insecticide. A decoction from leaves is reported to be a useful febrifuge (Soukup, Martinez). The species is recommended by Menninger for seaside gardens.

Ref.: Adams; Bircher; Burkill; Calabria; Chittenden; Kunkel 69; Little & Wadsworth; Long & Lakela; Martinez; Menninger 64; Mortensen & Bullard; Neal; Pesman; Purseglove; Soukup; Uphof.

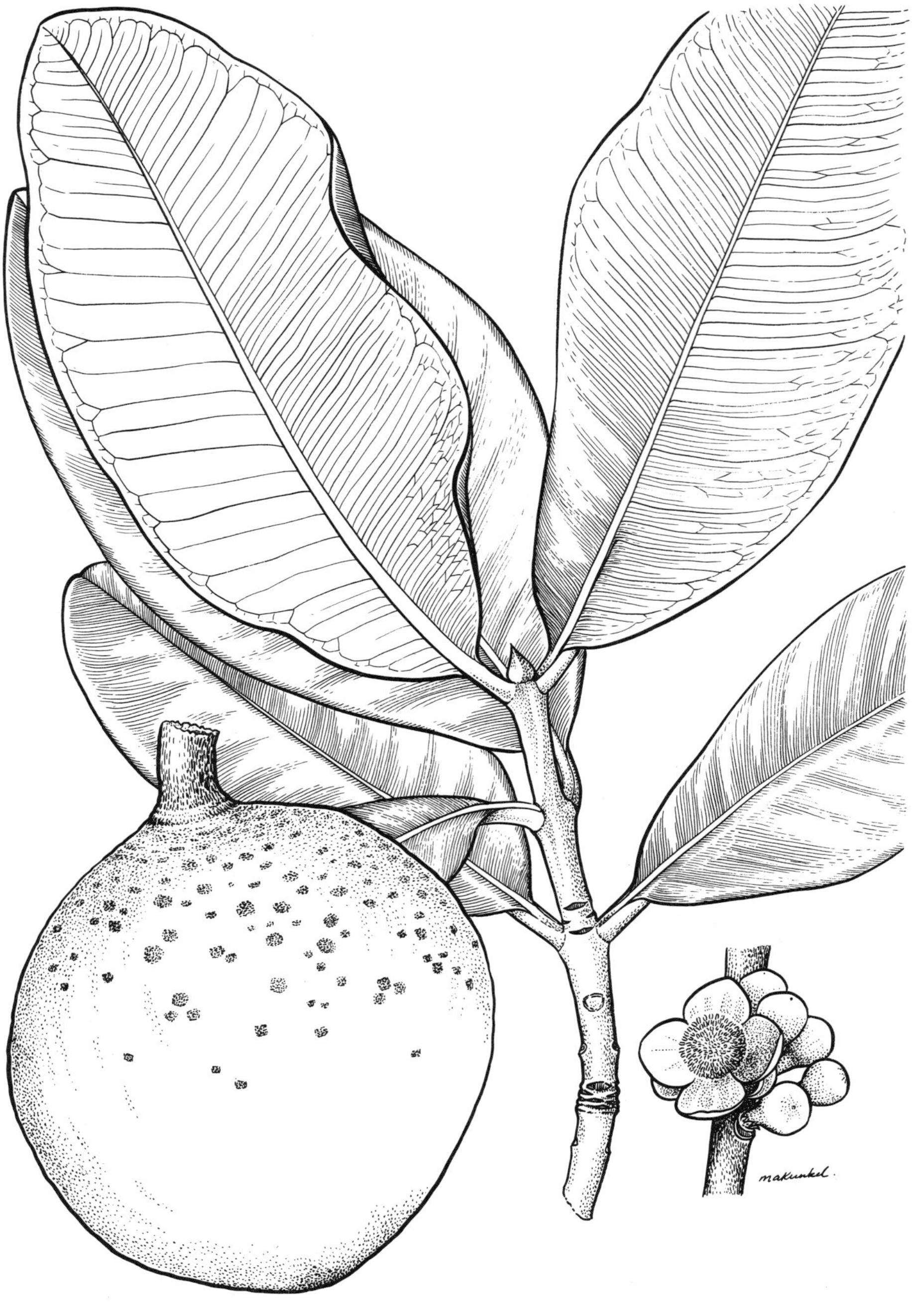

Fig. 41. *Mammea americana*; leaves, fruit and flowers = 3/5

FLACOURTIACEAE, the Flacourtia family

This family, which is chiefly native in the tropics, consists of some 90 genera and 1.000 species, mostly shrubs or trees. Some are of high ornamental value, others are appreciated for their edible fruits. However there are relatively few species in cultivation. One of these is the Governor's Plum (*Flacourtia indica*), a native of tropical Africa and Asia, the fruits of which are most delicious but it is hardly ever seen in these islands. Also rare but recorded as present is

Dovyalis hebecarpa (Gardn.) Warb. – Ceylon Gooseberry

A shrub or small spreading tree native in Sri Lanka (Ceylon). It may reach 6 to 7 m in height, and branches of older specimens are said to be armed with stout thorns. Specimens found cultivated in these islands are unarmed and much smaller. The Ceylon Gooseberry is supposed to be evergreen, having stalked oval-shaped, glabrous or slightly hairy leaves 6 to 8 cm long, tapering towards the apex, with a remarkably undulate or even curled-up appearance. Flowers small and inconspicuous, of a greenish colour; they might be uni- or bisexual. Of some interest are the globular fleshy fruits up to 2,5 cm in diameter borne on short stalks. These fruits are purple-coloured, have a velvety skin, and are very acid indeed; however they may be made into jam or jelly if enough sugar is added.

According to Burkill the native name of this species is 'Ketambilla' and it grows in Ceylon from 1.300 m down to almost sea-level. Propagation is recommended by seeds, or from cuttings or layers. Related species cultivated in other countries include the Kei Apple (*Dovyalis caffra*), from South Africa, and probably an East African species, *D.abyssinica*, both with less acid fruits.

Ref.: Bircher; Burkill; (Chittenden); Graf; Kennard & Winters; Little & al.; Mortensen & Bullard; Neal; Uphof.

Fig. 42. *Dovyalis hebecarpa*; fruiting branch = 3/5

SALICACEAE, the Willow family

Apparently 3 genera with some 530 species of shrubs and trees (Willis) mainly native in temperate zones of the Northern Hemisphere (Long & Lakela: 2 gen., about 180 spp.). Most species (approx. 500) belong to the genus *Salix* which has produced numerous hybrids so adding to the existing taxonomic confusion within this plant group. The members of the family have unisexual flowers which appear on separate trees (dioecious). Some Poplars (*Populus*) and Willows (*Salix*) are of economic importance, others serve for reafforestation purposes. *Salix canariensis* Chr.Sm. ex Link is a native of the Canary Islands. Of the introduced species the following are most frequent:

Populus alba Linné – White Poplar

Obviously not as common in these islands as originally believed as most specimens seen belong to the closely related Grey Poplar cited below. The true White Poplar is a tree up to 25 m in height, often branched right from the base, with abundant rootshoots or suckers. Crown spreading; bark of branches and younger stems whitish, twigs tomentose. Foliage deciduous. Mature leaves long-stalked, approximately triangular in outline, 3 to 5-lobed and coarsely toothed; blade subcoriaceous, up to 8 cm long and wide, dark green and glabrous above and whitish-tomentose beneath; young leaves tomentose all over. Fruiting catkins 5 to 8 cm long.
A species of humid localities where it may become aggressive. Sometimes seen as an ornamental. Propagation from cuttings and suckers. The White Poplar is thought to be native in Central and Eastern Europe, and is naturalized now in most Mediterranean countries. Several varieties known. Wood used to make shoes, matches, packing-cases, cellulose etc. (Uphof); the bark has medicinal applications (tonic, febrifuge).

Ref.: Chanes; Chittenden; El Hadidi & Boulos; Encke; Fitschen; Franco 64; Hart & Raymond; Lombardo 58; Mitchell; Ruiz de la Torre; Uphof.

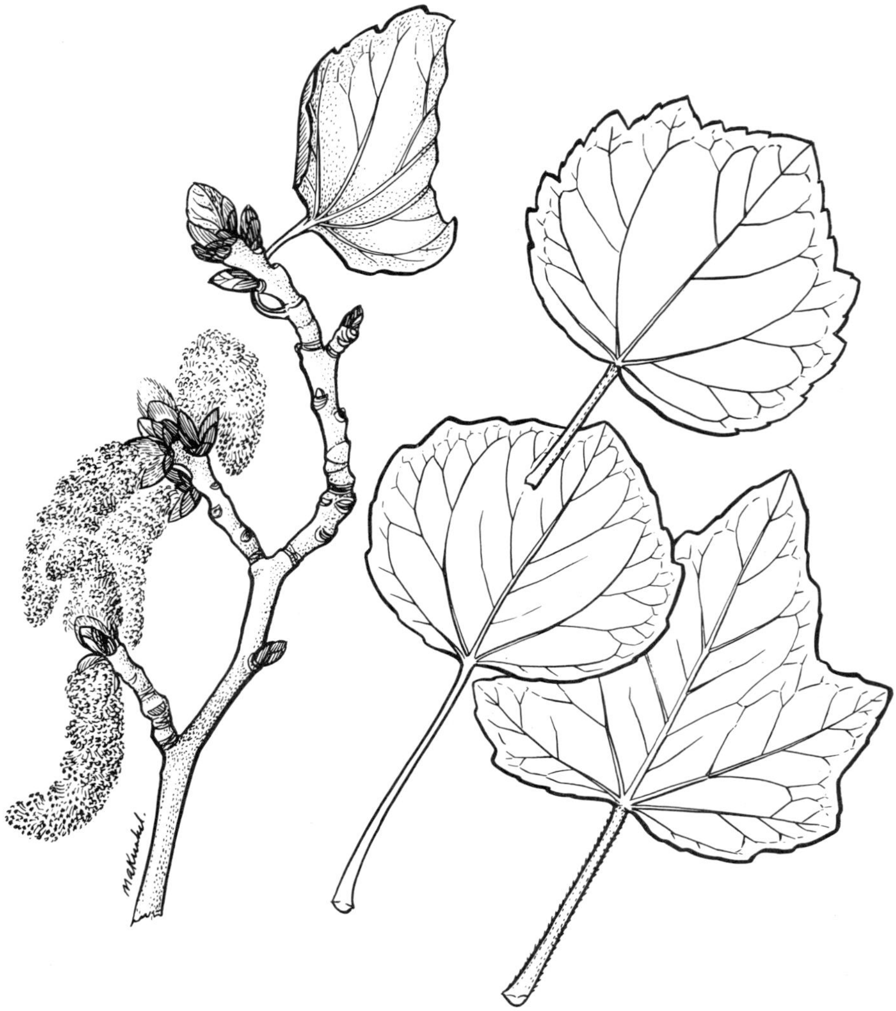

Fig. 43. *Populus canescens*; flowering branch and leaves (centre), the more triangular leaf (right) belongs to *P.alba*; all drawings about 3/5

Populus canescens (Ait.) Sm. – Grey Poplar

Closely related to the above and often confused with it, this rather fast growing species belongs to the southern European region. It is distinguished from the White Poplar by the cordate-deltate leaves which are greyish-tomentose beneath; petiole longer, thinner and pale yellow, the leaf-margin is shallowly dentate to somewhat sinuate. Blade of mature leaves coriaceous, usually up to 5 cm long and 6 cm wide; leaves of suckers confusingly similar to those of the White Poplar. Bark dark grey, rather corky.
Although cited in 'Flora Europaea' (Franco) as a *species* it is suggested (Fitschen, Mitchell) that the Grey Poplar could be an old intermediate or hybrid form between *Populus alba* and *P.tremula* (the Aspen). It resembles the former by the tomentum of its leaves, and the latter by the long thin petiole, leaf-shape, and the way in which the leaves move and rustle even on calm days.

Ref.: Bircher; Chittenden; Fitschen; Franco 64; Lombardo 58; Mitchell.

Populus nigra Linné – Black Poplar

A more common and quite different species often found along roadsides, riverbeds and waterchannels. The typical form of this species (native in most parts of Europe) is a tree with an upright or spreading crown. More frequent however is the so-called Lombardy Poplar: *Populus nigra* cv. 'Italica' (*P.pyramidalis* Rozan.), which is easily recognized by its columnar growth habit.
Tree 15 to 25 m tall, trunk usually very short, basal shoots frequent. Bark dark grey, somewhat furrowed in older specimens. Branches erect, rather flexible. Foliage deciduous. Leaves long-stalked; blade subcoriaceous, dark green, glabrous, triangular to almost rhomboid, up to 8 by 8 cm; margin serrate. Fruiting catkins 10 to 15 cm long.
Easily propagated from cuttings and basal shoots. A fast growing tree suitable for cooler climates; less successful for reafforestation except in moister soils. According to Font Quer a decoction of the buds and resinous substances is used to cure haemorrhoids; it is 'also considered an excellent medium for hair-growing' (Uphof). Wood light, used for packing-cases and paper pulp. – In cultivation also *Populus × canadensis* Moench (*P. × euramericana*), a very variable, probably triple-hybrid of *P.nigra × deltoides × angulata* which is best considered with its numerous cultivars (see Franco, Chittenden, Mitchell).

Ref.: Bircher; Chanes; Chittenden; (El Hadidi & Boulos); Encke; Fitschen; Font Quer; Franco 64; Harrison & Raymond; Lombardo 58; Mitchell; Ruiz de la Torre; Uphof.

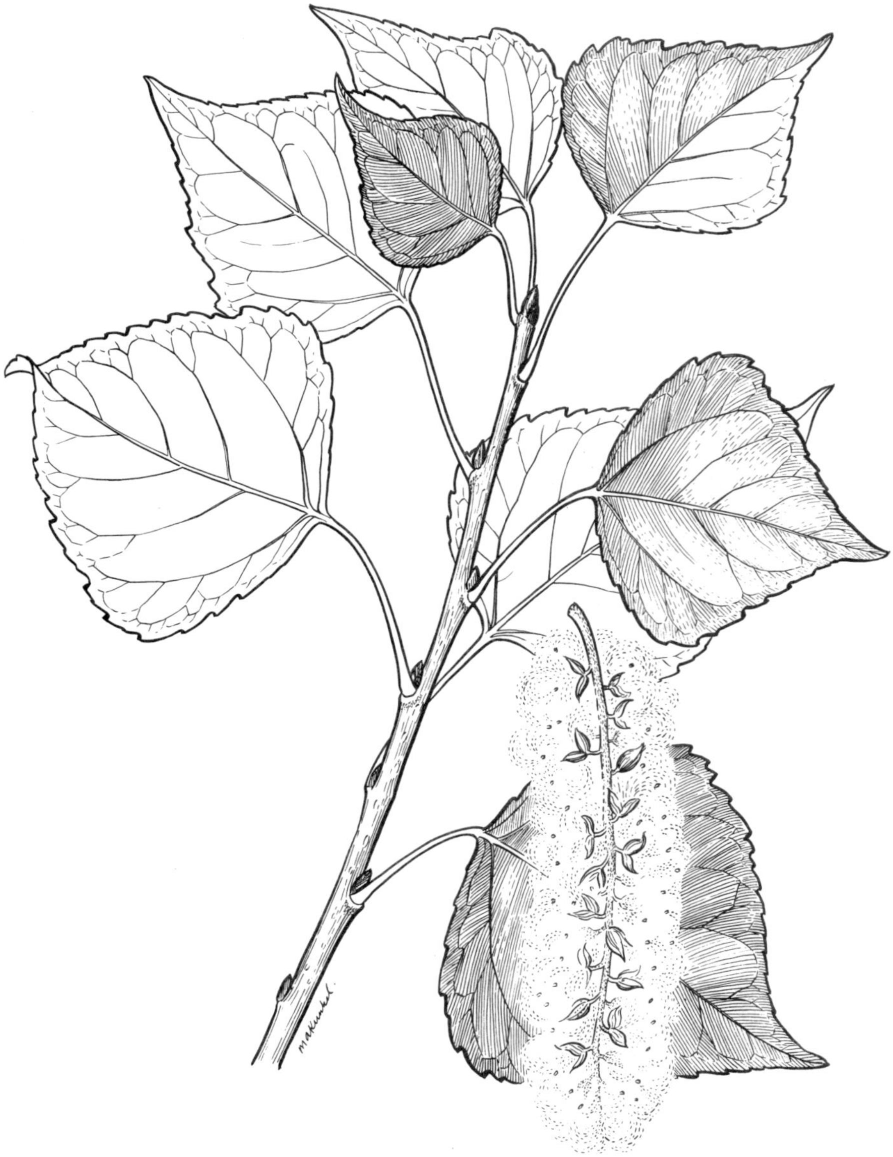

Fig. 44. *Populus nigra* cv. 'Italica'; drawing = 3/5

Salix spp. – Willows

K.-H. Rechinger in 'Flora Europaea' deals with 3 subgenera, 70 'good' species, about 10 subspecies and some 25 hybrids known as wild-growing in Europe. It needs only *one* specialist to sort out this labyrinth because two or three of them might never be 'd'accord'. In these islands only 3 species to be briefly considered, all of which are closely related and hybridize almost too freely for our nomenclatural convenience.

Salix alba Linné – White Willow

This rather confusing complex species (several subspecies and hybrids) is seen locally only of weeping habit and is commonly mistaken for the true or Chinese Weeping Willow as cited below. Small to medium-sized tree with a short but pronounced trunk, bark greyish-brown, irregularly fissured. Crown rounded and – as our description hopefully refers to cv. 'Tristis' of the subspecies *vitellina* (L.) Arc. – with arching, then drooping terminal branches. Twigs very flexible, orange-coloured. Foliage of well-watered garden specimens semi-persistant. Leaves short-stalked, lanceolate, up to 10 cm long and 1,5 cm wide; herbaceous to subcoriaceous, finely serrate, pale green and somewhat shiny above, and slightly tomentose, glaucous-white beneath. Catkins axillary, cylindrical, the fruiting ones up to 8 cm long.
Willows are usually propagated from cuttings. They grow well in a moist, cooler environment. The Weeping Willow (as this form is called) is a gracious tree along riversides and in freely watered gardens. – For its virtues as febrifuge and anti-rheumatic see also Font Quer. According to Uphof the 'leaves are employed as substitute for tea'. Wood used in carpentry, for tool-handles and paper pulp.

Ref.: Chanes; (Chittenden); Encke; Fitschen; Font Quer; (Harrison); (Lombardo 58?); Mitchell; Polunin; Rechinger; Ruiz de la Torre; Uphof.

Near related and often confused with the pendulous form of *Salix alba* is the true or Chinese Weeping Willow – *Salix babylonica* L. – which, in spite of its misleading botanical name is supposed to be native in China. The mature leaves of this species are linear-lanceolate, up to 15 cm long and almost glabrous on both sides. *Salix fragilis* L., the Crack Willow (from Eurasia), is cultivated for its workable twigs which are made into baskets; it may be recognized by its frequently pruned appearance, the pale twigs, and the larger leaves up to 15 cm long and 4 cm wide.

Fig. 45. *Salix alba*; drawing and details = 3/5

ERICACEAE, the Heather family

About 50–55 genera, with 1.300 to 1.500 species; mainly shrubs and some small trees. From cooler to subtropical climates, with main distribution centres in South Africa and East Asia. Most important genera *Erica* (about 500 species) and *Rhododendron* (+ *Azalea*) with perhaps 600 species. The family provides numerous fine ornamentals, including the usually poisonous Rhododendrons; *Arbutus* and *Vaccinium* have edible fruits. *Arbutus canariensis* is a Canarian endemic; *Erica arborea* and *E.scoparia* are Canarian natives but have a wider distribution.

Arbutus unedo Linné – Strawberry Tree

Also known as Madroño, this species is a native of the Mediterranean region and – curiously enough – Southwest Ireland; it hybridizes with *A.andrachne* L. (from the eastern Mediterranean, especially Greece). Evergreen tree 8 to 10 m tall; crown dense and spreading. Trunk short, up to 40 cm in diameter, often with basal shoots; young branches reddish. Bark reddish-brown turning grey in older specimens, otherwise scaly and flaking off. Leaves alternate, stalked, oblong-lanceolate, coriaceous, dark green (new leaves shiny), 5 to 8 cm long and up to 2,5 cm wide; margin dentate. Leaf-stalks slightly tomentose. Flowers white to pinkish or tinged with green; corolla urceolate or pitcher-shaped, less than 1 cm in diameter; in showy, drooping terminal panicles. Fruit a rounded berry up to 2 cm in diameter, scarlet to crimson when ripe, with a very rough surface; numerous small seeds surrounded by the edible pulp.

Propagated from seeds and cuttings (tips), and best grown in a somewhat cooler climate. The fruits are sweet but mealy and are made into preserves and alcoholic beverages. Leaves astringent, bark used as an antiseptic; leaves and bark applied for tanning. Wood heavy, used for cabinet-making; also for fuel and charcoal. Otherwise a good shade tree and a fine ornamental for larger gardens. Several varieties of this species are described by Bean and others.

Ref.: Bean; Chittenden; Eliovson; Elwes & Henry; Encke; Font Quer; Graf; Harrison; Kunkel 69; Lombardo 61; Menninger 64; Mitchell; Perry; Polunin; Polunin & Huxley; Polunin & Smythies; Ruiz de la Torre; Uphof; Webb.

Fig. 46. *Arbutus unedo*; drawing and details = 4/5

Erica arborea Linné – Tree Heath

Usually seen as a shrub this species may grow into a fine tree up to 15 (20) m in height, or be even taller in dense forests where it may attain an impressive trunk up to 80(!) cm in diameter. Bark dark grey-brown, thin-scaly or flaky. Crown spreading in well-developed specimens; in bushy specimens crown narrow, more or less columnar. Evergreen but continually shedding old leaves. Leaves needle-like, dark green, rather hard, 6 to 8 mm long. Flowers small, whitish, almost bell-shaped, 2 mm in diameter, more or less terminal on new branchlets which are slightly downy. Seeds minute.
Although a species native in these islands, it is cultivated elsewhere and therefore included in this account. Propagation from seeds; perhaps also from cuttings. The Tree Heath (locally 'Brezo') makes a good hedge and is useful for reafforestation on eroded slopes in humid subtropics but declared a noxious weed in New Zealand (Parham & Healy). Wood hard, used for manufacturing bruyère pipes, bowls and other articles. – This species has a curious, rather disjunctive distribution: it is native in the Atlantic Islands, parts of the Mediterranean region and SW Europe, the Ethiopian mountains and, according to Palmer & Pitman, in the 'Tibetsi (sic!) Mountains in the heart of the Sahara'.

Ref.: Bean; Chanes; Chittenden; Encke; Kunkel & Kunkel; Palmer & Pitman; Parham & Healy; Perry; Polunin; Polunin & Huxley; Polunin & Smythies; Ruiz de la Torre; Uphof; Webb & Rix.

Fig. 47. *Erica arborea*; natural size

EBENACEAE, the Ebony family

Three or four genera with over 500 species of tropical trees and shrubs of which the species of *Diospyros* are especially appreciated as valuable timber trees; they are the source of the dark, heavy and very hard Ebony wood. One species widely cultivated as a fruit tree:

Diospyros kaki L.f. – Japanese Persimmon

Deciduous tree native in East Asia; commercially grown in higher regions of Mediterranean climates. Up to 12 m tall; trunk short, bark greyish and fissured or furrowed. Dense,dark and spreading crown while in leaf; branches often somewhat pendulous. Leaves alternate, petiolate, broad oval-shaped, herbaceous, up to 10 cm long and 6 cm broad; leaves of young plants much larger. Margin somewhat undulate. Very showy in the autumn as the leaves turn colour to an almost startling red-orange. Pollen-bearing flowers in groups up to three; fruit-yielding flowers solitary, of varying greenish colours. Fruit globose, slightly flattened (resembling a tomato), up to 7 cm in diameter, thin-skinned, orange-red or yellowish; the fruit is juicy and edible having a delicate flavour hard to define. Seeds oblong, flattened and embedded in the pulp.

There are numerous cultivars of this species. The fruits are eaten raw or consumed when dry. Unripe fruits astringent. Tannin of unripe fruits used for dyeing and for preserving wood. The flowers are applied in the treatment of coughs. The hard wood is sometimes employed in local carpentry. – Alternative names: Kaki, Kakee, Chinese Persimmon, and Japanese Date plum.

Ref.: Bean; Bircher; Carbo & Vidal; Chanes; Chittenden; Eliovson; Encke; Esdorn & Pirson; Fitschen; Harrison; Kunkel 69; Marzocca 50; Mitchell; Mortensen & Bullard; Neal; Uphof.

A second species, *Diospyros lotus* L. (the Date Plum, also edible), has been tried in these islands but given up as a failure. It is now recommended as a stock for the above described Persimmon.

Fig. 48. *Diospyros kaki*; drawing = 3/5

SAPOTACEAE, the Sapodilla family

A large tropical family of trees and shrubs; usually evergreen, with milky sap. According to Neal about 30 genera and 400 species, to Rendle 40 genera and 600 species, and to Willis '35–75 ill-defined' genera with 800 species, which means it is 'a rather modern family characterized by inflation'. – The family includes a number of plants of economic importance such as *Manilkara achras* (Mill.) Fosberg (the Chicle tree, from Central America), *Chrysophyllum cainito* L. (the Star Apple, from C.America), *Palaquium gutta* (Hook.) Burck. (the Gutta-percha tree, from Indomalaysia), *Butyrospermum paradoxum* (Gaertn.f.) Hepper ssp. *parkii* (the Shea-butter tree, from tropical Africa), etc., none of them successfully established in Canary gardens. Another species – *Sideroxylum marmulano* Banks ex Lowe – is native in the Macaronesian region.

Pouteria campechiana (H.B.K.) Baehni – Egg-fruit Tree

Also known as 'Canistel', or under its synonyms *Lucuma nervosa* A.DC. and *L.rivicoa* Gaertn.f., this fruit tree is a native of northern South America. Evergreen, 8 to 10 m tall, with a dense, dark, more or less pyramidal crown. Trunk short, up to 40 cm in diameter, frequently with basal shoots. Bark dark grey and fissured, scaly in older specimens. Leaves alternate, petiolate, dark green and glabrous; blade narrow-oblong or oblanceolate, 12 to 15(20) cm long and 4 to 6 cm wide, with strong lateral veins; margin slightly undulate. Flowers solitary or clustered, axillary, stalked, greenish and little conspicuous. Fruit large and 'fleshy', somewhat mango-shaped but pointed, up to 6 cm in diameter, with a thin orange-yellow skin covering a soft, mealy, musky, 'hard-boiled-egg-yolk-coloured' pulp. One to three dark, shiny and boat-shaped seeds.

Fruits are eaten raw (not everyone's taste!), or cooked and prepared in custards and ice-cream. According to Barrett the pulp is high in carbohydrates. The species is propagated from seeds and grows best in warm moist climates of lower altitudes. – The species is recommended to the R.S.P.C.A. which is much against overcrowded egg-producing chicken farms.

Ref.: Barrett 56; (Bircher); Blackwell; (Burkill); (Chittenden); Kunkel 69; Long & Lakela; (Menninger 64); Neal; (Purseglove); (Uphof).

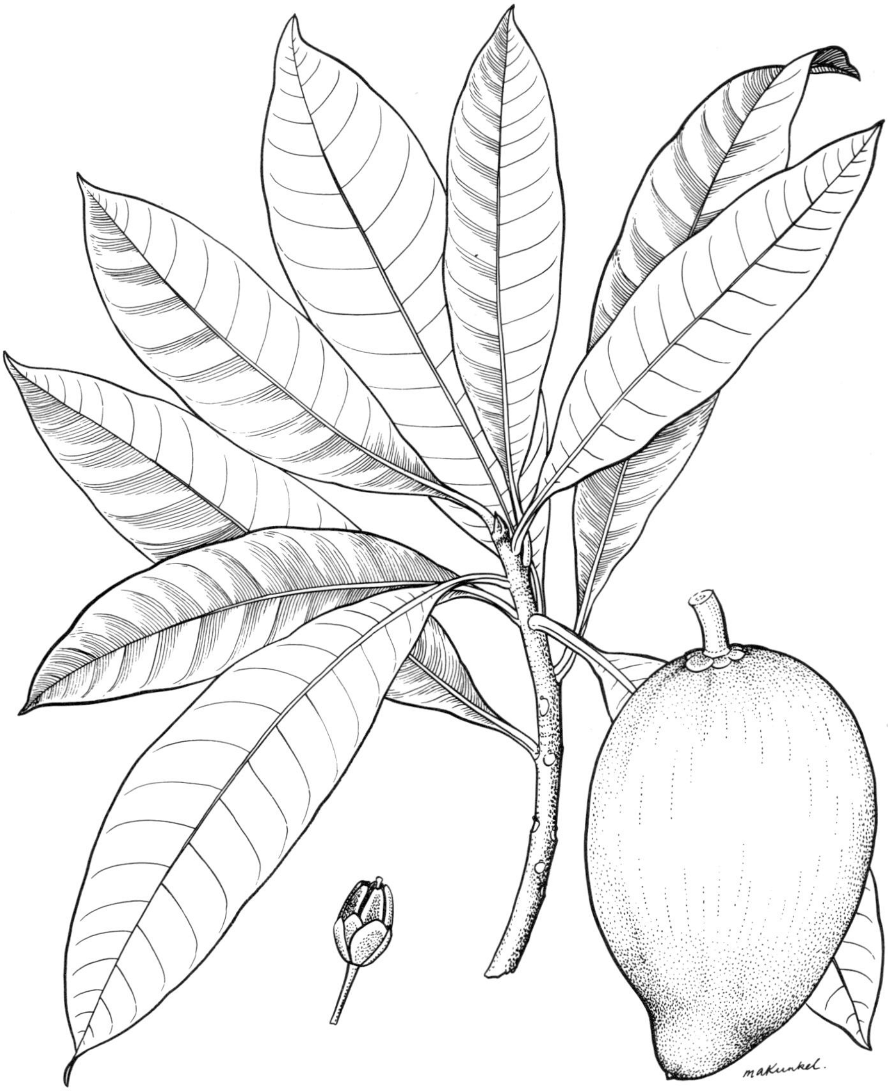

Fig. 49. *Pouteria campechiana*; drawing and details = 3/5

TILIACEAE, the Lime or Linden family

About 45 to 50 genera with almost 500 species of trees and shrubs (few herbs), chiefly from Southeast Asia and South America. *Corchorus* produces Jute, and *Tilia* (Northern Hemisphere) provides some good timber trees; there is no native Tiliaceae in these islands.

Tilia platyphyllos Scopoli – Large-leaved Lime

Tree up to 35 m tall; in the Canary Islands usually 8 to 12 meters. Trunk straight, up to 1 m in diameter; bark greyish and fissured. Crown spreading or rounded; branchlets somewhat tomentose, with pronounced axillary buds. Foliage deciduous. Leaves alternate, long-stalked, more or less pendent; blade heart-shaped, herbaceous, mid-green, pubescent beneath, up to 10 cm long and wide; with a sharply dentate-serrate margin and a pronounced nerve system. Flowers whitish, in 2 to 5-flowered, long-stalked, axillary inflorescences accompanied by a partly detached laminar bract up to 6 cm long. Fruits dry, globose but ribbed, indehiscent, tomentose; usually 1-seeded. Propagation from cuttings. A somewhat variable species and best grown in cooler climates. Wood used in homestead carpentry, and the fibre of the bark for ropes and mats. Most important are the flowers of which an infusion is made as a tea and mouth-wash, for stomach aches and to calm nervous tension. The species is native in most countries of central and southern Europe, and is usually planted in parks and as a roadside tree. – Several subspecies and varieties are known. There are also a few specimens of *Tilia tomentosa* Moench (Silver Lime) and of the large-leaved American Lime (*T.americana* L.) in Canary gardens.

Ref.: Browicz; Chanes; Chittenden; Dimitri & Alberti; Encke; Fitschen; Font Quer; (Harrison); Lombardo 58; Mitchell; Ruiz de la Torre.

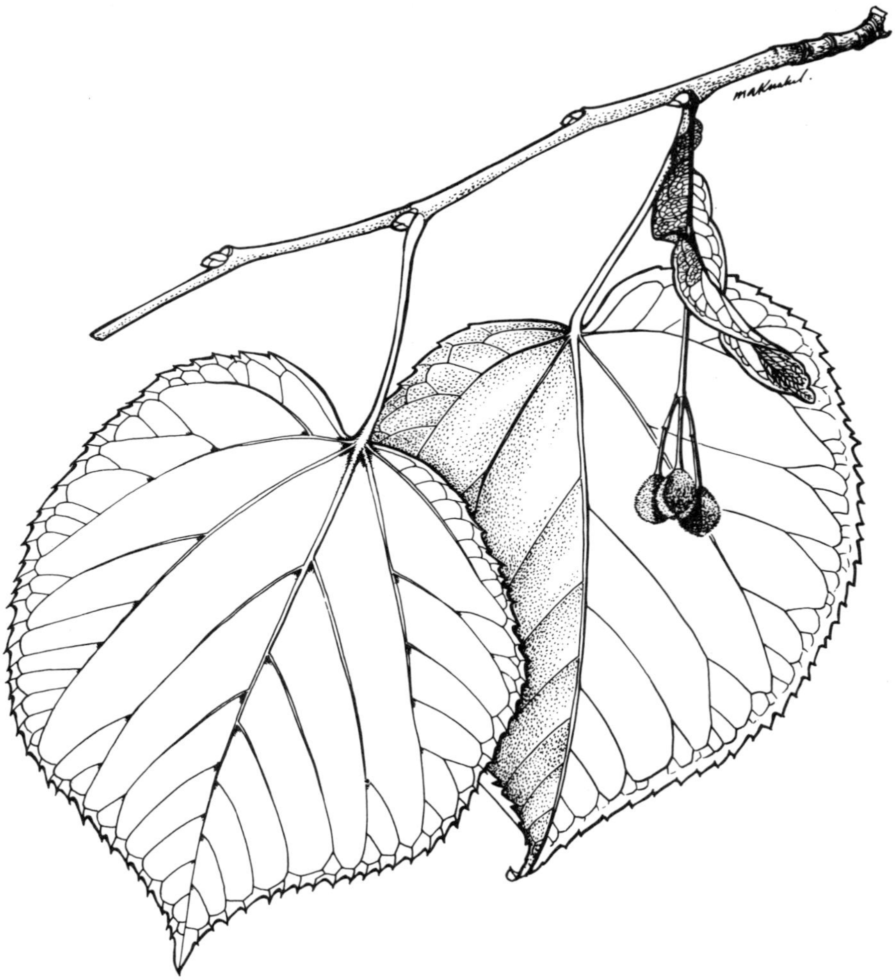

Fig. 50. *Tilia platyphyllos*; fruiting branch = 1/2

STERCULIACEAE, the Cocoa family

This family of mainly tropical trees, shrubs, vines and some herbs is composed of about 50 genera with more than 700 species. Commercially the most important species is *Theobroma cacao* L., the Cocoa tree, from tropical America. *Tarrietia, Triplochiton* and others are valuable timber trees. Some plants (such as *Abroma*) yield a fibre, others have medicinal properties (e.g. *Waltheria*) or provide the Coca Cola factories with a stimulant for a soft drink (*Cola*), whereas a few species are planted just because they are beautiful.

Brachychiton acerifolium (A.Cunn.) F.v.Muell. – Illawarra Flame

An ornamental medium-sized or tall tree from Australia, often also called the 'Flame Bottle-tree'. Trunk straight, frequently somewhat swollen, up to 60 (or 80) cm in diameter; crown more or less pyramidal. Bark dark grey or greyish-green, fissured. Briefly deciduous. Leaves alternate, long-stalked, leathery, deep green and shiny; blade up to 20 cm long but rather polymorphic, usually three-lobed, sometimes entire or up to 5-lobed. Twigs with leaf-scars. Flowers bell-shaped, 1,5 to 2 cm in diameter, bright red, in large, rather spectacular terminal or axillary panicles. Fruit-capsule woody, black, boat-shaped when open; up to 15 cm long, on stalks of almost the same length. Seeds 8 to 10 mm long, yellow, covered by a soft, brownish tomentose bedding.

Propagation from seeds and cuttings. A good park and roadside tree growing best at lower altitudes. It is especially showy when in full flower after shedding most of its leaves. The bark exudes a gum, and yields fibre which is used for making ropes. – It is the 'Flame-tree' of Eliovson who reports the species as almost evergreen. Hybridization with the following or other species seems to occur.

Ref.: Barrett 56; Bircher; (Chittenden); Eliovson; Graf; (Harrison); Kunkel 69; Lombardo 58; Menninger 62; Molinari & Milano; Neal.

Fig. 51. *Brachychiton acerifolium*; leaves, inflororescence and fruits = 2/5, solitary flower = 1/2

Brachychiton diversifolium (Don) R.Br. – Kurrajong

The related Kurrajong also makes a fine park tree although it remains smaller and has less showy flowers. It is an evergreen species and has leaves similar to those of the common black poplar, reason for which it is often cited as *B.populneum* R.Br. Leaves long petioled, blades 6 to 8 cm long and up to 4 cm wide, with a pronouncedly long and narrow tip. Flowers bell-shaped, greenish-white, up to 1 cm in diameter; fruit capsules dark brown, woody, boat-shaped, 5 to 7 cm long. Seeds yellow, partly covered by rigid hairs.

Propagated from seeds and cuttings. It thrives best from near to sea-level up to some 500 meters, and makes an excellent shade tree. The Kurrajong is also native in Australia. According to Bircher its bark produces a very tough bast and still serves for clothing in some areas; the leaves are said to be eaten by animals. El Hadidi & Boulos call this species simply the 'Bottle tree'.

Ref.: Barrett 56; (Bircher); (Chanes); (Chittenden); (El Hadidi & Boulos); (Eliovson); (Harrison); Kunkel 69; (Lombardo 58); Menninger 62; (Moeller 71); (Molinari & Milano); Neal.

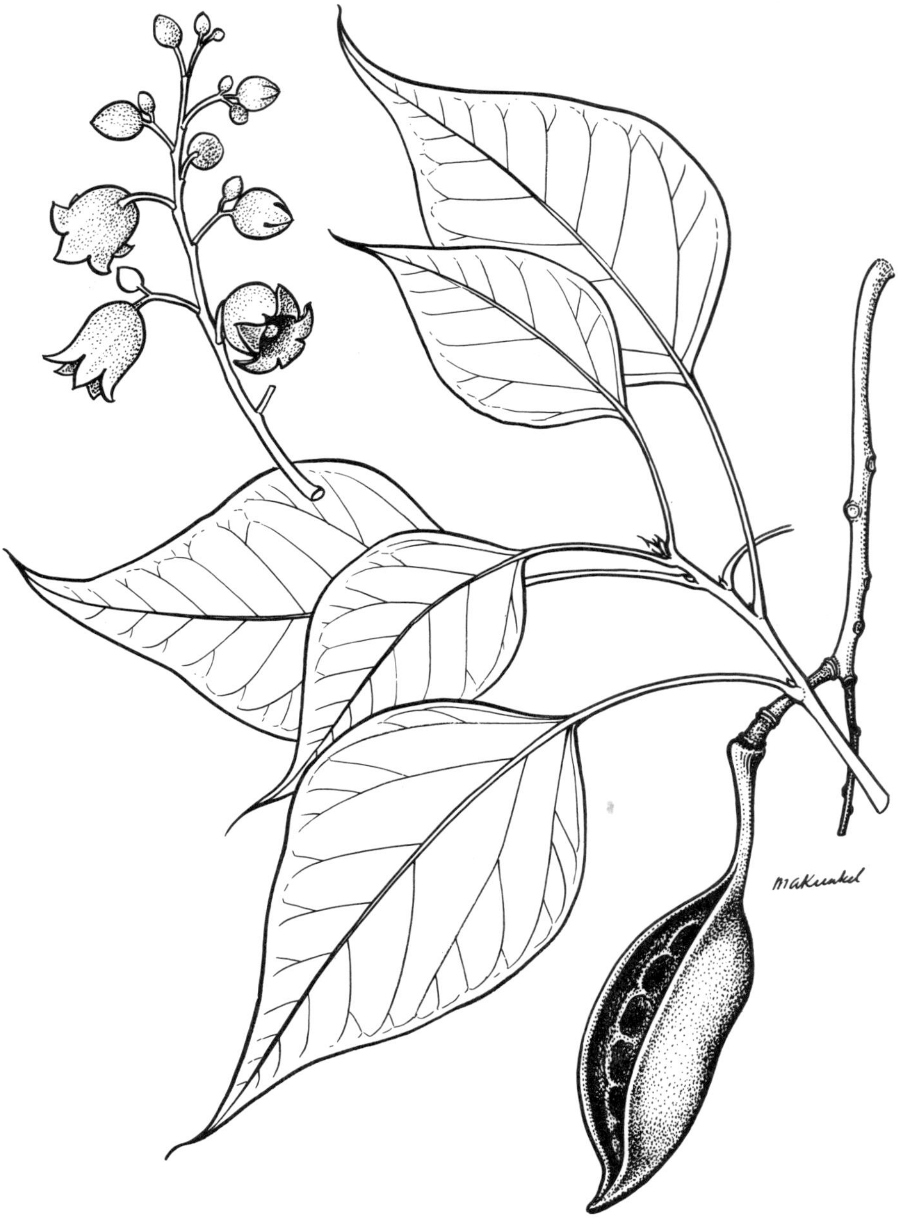

Fig. 52. *Brachychiton diversifolium*; drawings = 3/5

Dombeya × *cayeuxii* hort. ex André – Pink Ball

This species is considered to be a hybrid between *Dombeya mastersii* and *D.wallichii* but is often (wrongly) cited under the latter name. The mentioned parents are supposed to be native in Madagascar and tropical East Africa (Barrett).

Usually a large shrub and tree-shaped only when frequently pruned, the Pink Ball reaches 6 to 7 meters in height. Trunk short, knotty, with a brownish-grey, fissured bark. Crown dense and almost rounded; twigs pubescent. Apparently evergreen. Leaves alternate, with petioles up to 12 cm long. Blade very soft and hairy, almost rounded or angular towards the apex, and cordate and lobed towards the base, dark green, about 20 cm wide. Flowers pink (brown when old), in long-stalked, globose, very dense clusters; rather persistant. Fruits dry, light brown capsules up to 1,5 cm in diameter.

Propagation from cuttings; it grows best in more temperate climates. It makes a beautiful garden and street tree and is a sight to be remembered around Christmas whilst in full bloom.

Ref.: Adams; Barrett 56; (Beyron); (Bircher); Chittenden; Encke; (Graf); Kunkel 69; (Lombardo 61); (Moeller 68); (Molinari & Milano); (Neal); (Schaeffer).

Fig. 53. *Dombeya* × *cayeuxii*; flowering branch and solitary fruit = 2/5

A second species of the genus and only recently introduced is known as

Dombeya tiliacea (E.Mey. ex Endl.) Planch. – Forest Dog Rose

Although in these islands still a shrub, it is said to become a small tree up to 9 m in height and could have a trunk of 30 cm in diameter (Palmer & Pitman). This little tree is a native of South Africa.

Branches slender and spreading; foliage briefly deciduous, leaves stalked, heart-shaped or lobed resembling those of the grape vine but are soft and have minute star-shaped hairs. Flowers axillary or terminal, solitary or in clusters, white turning brownish when drying, up to 3,5 cm across and rather persistant. Fruit capsule pale brown, rounded, 4 to 7 mm in diameter and protected by the persistant petals. Seeds black. – The Forest Dog Rose, as far as we know, is only of ornamental value. It is grown from seeds but may also strike from cuttings.

Ref.: Neal?; Palmer & Pitman; Van der Spuy.

Fig. 54. *Dombeya tiliacea*; drawings = 3/5

BOMBACACEAE, the Cotton-tree family

A family of about 20 genera and 180 species, chiefly large trees of tropical countries. Several species thrive well in drier climates, and some are appreciated ornamentals for larger gardens.

Chorisia speciosa St.Hill – Floss-silk Tree

A medium-sized to large tree from eastern South America, up to 20 m tall. Trunk columnar and slightly inflated ('bottle-tree'); bark green, shiny, and strongly armed by stout dark thorns. If pruned early the plant may produce several stems. Wood soft, and branches break off easily in strong winds. Deciduous. Leaves long-stalked, digitate-palmatisect, with 5 to 7 oblong or oblanceolate leaflets which are herbaceous, of a medium green colour, long-pointed at the apex; margin dentate. Single leaflets up to 12 cm long. Flowers extremely showy and often compared with those of certain orchids. They usually appear in pairs or small clusters and have a bell-shaped tubular calyx up to 2,5 cm long; petals subspathulate, slightly undulate or lacerate, 6 to 8 cm long; rose-coloured with purple streaks, and whitish towards the yellow centre. Stamens up to 8 cm long. A fully open flower may reach up to 15 cm in diameter; the tree flowers whilst leafless. Fruit-capsule oblong or somewhat pear-shaped, up to 15 cm long, with many small seeds embedded in a mass of silky hairs.
The species grows from seeds and from cuttings. It is cultivated in many countries and most impressive specimens are found in Madeira where an entire ravine is full of Floss-silk trees. – Best grown at lower elevations. Considering its remarkable flowers one can only agree with Little & al. in saying that 'this relative of ceiba is one of the world's most beautiful flowering trees'. Besides being ornamental, the silky hairs covering the seeds are used as pillow and cushion stuffing. Barrett records that 'the inner bark strips off in ribbons and may be twisted into cord for fishing nets'. According to Menninger several colour-forms are known.

Ref.: Barrett 56; Bircher; Chittenden; Eliovson; (El Hadidi & Boulos); Graf; Kunkel 69; Little & al.; Lombardo 58; Menninger 62; Neal; Uphof.

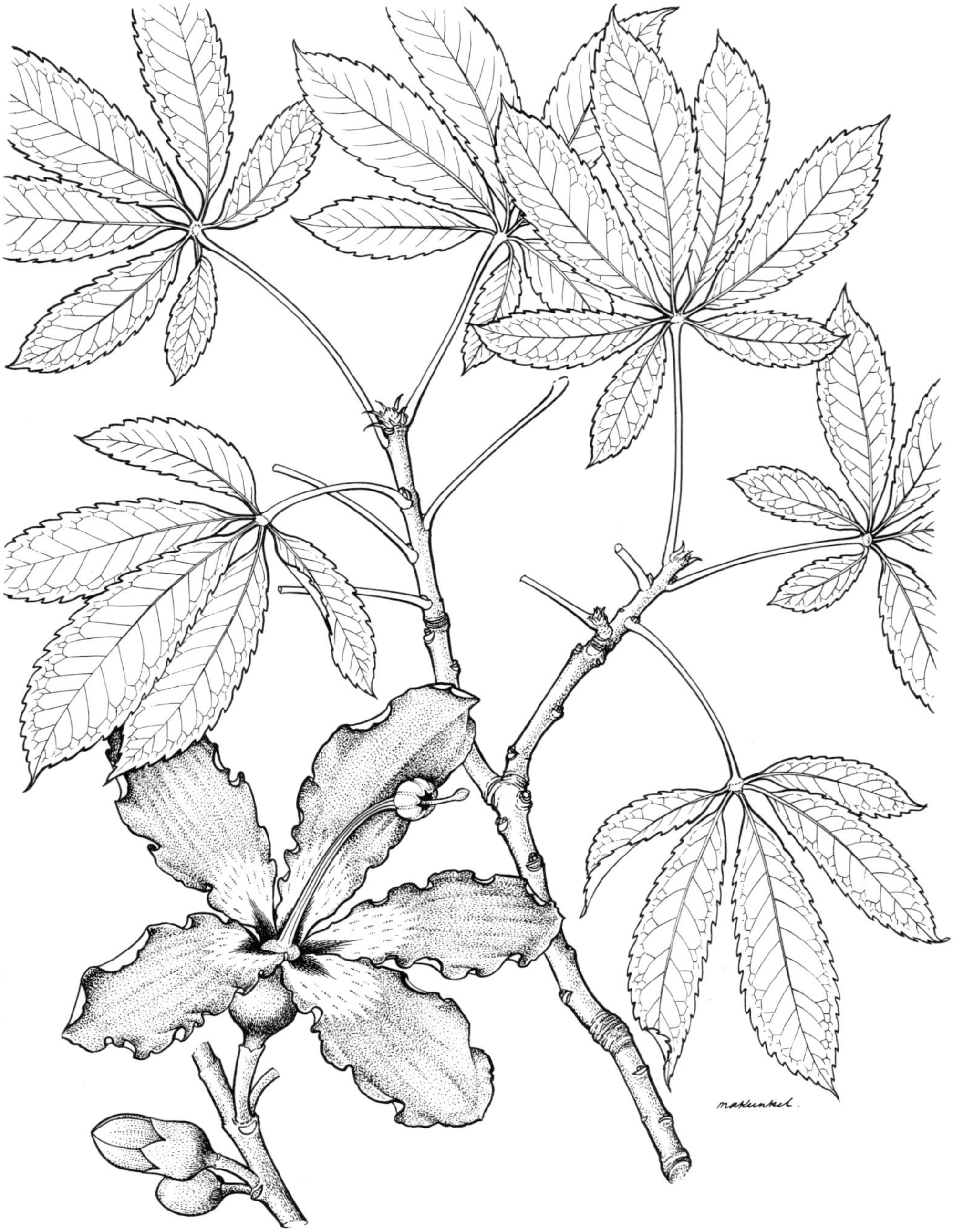

Fig. 55. *Chorisia speciosa*; drawings = 3/5

Pachira insignis (Sw.) Savigny – Shaving-brush Tree

An ornamental tree of which only few specimens are seen in our gardens. Native in the West Indies and northern South America; usually in somewhat drier regions. Other common names: Wild Breadnut, and Wild Chestnut. Small tree up to 10 m in height; trunk stout, with smooth greenish or greyish bark. Roots converting into buttresses. Branches in whorls, crown spreading. Evergreen (or very briefly deciduous). Leaves long-stalked, digitate, with 5 to 8 obovate leaflets which are tough herbaceous and up to 20 cm long; margin usually entire. Flowers solitary or in pairs on short stout stalks; petals narrow, up to 20 cm long, crimson, with a dense bunch of white and protruding stamens. Fruit capsule subglobose, brown, up to 20 cm in diameter.

Ref.: Barrett 56; Bircher; Chittenden; Graf; Little & al.; Menninger 62; (Neal).

Species of *Bombax* and *Ochroma* are rare in cultivation although some of the former genus are known to have very showy flowers, and the latter, with *O.pyramidale* (Cav.) Urban, is the source of Balsa, the lightest of commercial woods weighing less than cork. *Ceiba pentandra* (L.) Gaertn., the Silk-cotton tree whose natural occurrence in Africa and in Asian forests remains doubtful, is a very large buttressed tree of tropical America and not doing well in the Canary Islands. And missing too is the Baobab (*Adansonia digitata* L.), perhaps the most famous species of the entire family. Descriptions of the latter two species are given by Barrett, Irvine, Little & Wadsworth, Purseglove and other authors.

MALVACEAE, the Mallow family

A large family of herbs, shrubs and trees from temperate to tropical zones; about 75 genera and 1.000 species. *Abutilon*, *Althaea*, *Gossypium*, *Hibiscus*, *Malvaviscus*, *Pavonia* etc. are either ornamentals or grown for their commercial value; species of *Malva*, *Malvastrum* and *Sida* are cosmopolitan weeds. *Lavatera acerifolia* Cav. and *L.phoenicea* Vent. are beautiful flowering shrubs endemic to the Canary Islands.

Hibiscus elatus Swartz – Cuba Bast, or Mahoe

In gardens usually a small to medium-sized tree (8–12 m) but known to reach up to 25 m in Cuba and Jamaica, its home islands. Trunk straight, 25 to 40 cm in diameter; bark pale grey, smooth or finely fissured. Crown somewhat rounded; young branches with short hairs and ringed nodes. Evergreen. Leaves alternate, long-stalked, subcoriaceous and dark green; blade roughly heart-shaped or with a rounded base, pointed at apex; up to 12 cm long and wide. Flowers short-lived, grouped up to 3 in terminal position, yellow-orange in the morning turning red or purplish in the afternoon; petals narrow and somewhat curled. Fruit capsule ovoid, dark, tomentose, up to 3 cm in diameter, with many hairy seeds.

Propagation from seeds, sometimes also successful from cuttings. The species makes a fine ornamental; best grown at lower altitudes. The wood is used for cabinet work, furniture, gunstocks, shingles etc. Alternative common names: Blue Mahoe, Cuba Bark, and Mountain Mahoe. – The related *Hibiscus tiliaceus* L. is a more shrubby species which grows in swamps and by seashores; it has broad petals, a pointed seed-capsule, and the leaves are densely hairy beneath.

Ref.: Adams; Barrett 56; Bircher; Burkill; Kunkel 69; Little & al.; Menninger 62, 64; Neal; Uphof.

Fig. 56. *Hibiscus elatus*; flowering branch = 3/5

Lagunaria patersonii G.Don – White Wood

An Australian species, from Queensland, richly flowering and frequent in subtropical parks and gardens; also a roadside tree. Up to 15 m tall, with an almost pyramidal crown. Trunk straight, 30 to 50 cm in diameter; bark dark grey and fissured. Foliage evergreen. Leaves alternate, stalked, oblong-elliptic, coriaceous, dark green and rough above and very pale-tomentose beneath; 4 to 6 cm long and 2,5 to 3,5 cm wide. Flowers solitary or in groups towards the end of branchlets, quite showy; petals pink. Fruit capsule globose, quite hard, greyish, up to 3 cm in diameter. Seeds unfortunately surrounded by short rigid hairs which may cause skin diseases or other irritation, reason for its local name 'pica pica'.

Propagated from seeds and cuttings, the species grows in coastal zones as well as in drier inland areas but will not stand frost. The wood is used for constructions, and the bark yields a good fibre (Neal). The White Wood is admired in Canary gardens as it may flower for many months and is resistant to prolonged drought.

Ref.: Bircher; Chittenden; Eliovson; Graf; Harrison; Kunkel 69; Lombardo 58; Menninger 62; Molinari; Neal.

Fig. 57. *Lagunaria patersonii*; all drawings (incl. fruit) = 3/5

Thespesia populnea (L.) Sol. ex Correa – Portia Tree

The Portia Tree, Seaside Mahoe or Milo is said to be native in the eastern tropics and is naturalized in coastal thickets of tropical America and in West Africa where it is probably also introduced.
Evergreen tree up to 10 m in height, with a dense dark crown; trunk short and soon ramified. Bark pale grey, slightly fissured; branchlets often somewhat scaly. Leaves alternate, long-stalked, herbaceous to subcoriaceous, glossy dark green; blade narrow heart-shaped, up to 9 cm long and 7 cm wide. Flowers almost cup-like, single or in groups, more or less terminal, usually opening one at a time; petals broad, pale yellow turning purplish in the afternoon, so closing the 1-day flowering cycle. Fruit capsule yellowish-green, subglobose, indehiscent, up to 3 cm in diameter; seeds hairy.
This ornamental species is propagated from seeds and cuttings, and prefers warm and moist coastal climates. According to Uphof the wood is very hard and is used for boat-building, wheels and gunstocks; it is the source of the so-called Seychelles Rosewood. From the bark a string fibre is made. Irvine reports that leaves and fruits have medicinal properties, and Burkill states that 'throughout much of the Pacific the tree is sacred; it is planted about temples. This worship would seem to have come about on account of its great importance to sea-faring folk: that its leaves should be used in religious ceremonies followed'. – Mary F. Barrett's effort to translate the botanical name of every species dealt with into a 'pronounceable American' is to be admired, as in the case of *Thespesia* read the following: 'Thes-pee' she-a, or Thes-pee'se-a'.

An East African form of this species is described by Palmer & Pitman as *Thespesia acutiloba* (Bak.f.) Exell & Mendonça, and also said to be a cosmopolitan plant. It might be identical with *Th.populneoides* (Roxb.) Kostel. as discussed by Fosberg & Sachet.

Ref.: Adams; Barrett 56; Bircher; Burkill; Chittenden; Corner; Fosberg & Sachet; Irvine; Keay & al.; Kunkel 69; Little & Wadsworth; Long & Lakela; Menninger 62, 64; Neal; (Palmer & Pitman?); Uphof.

Fig. 58. *Thespesia populnea*; flowering branch and fruits = 3/5

EUPHORBIACEAE, the Spurge family

A very large family which consists of many genera and species representing practically all life forms in the Plant Kingdom; a family having caused much confusion and which will continue to give taxonomists many a headache until the last specimen of each species is properly provided with its own name. Only then may we know how many of the Euphorbiaceae exist. Rendle (1956) cites 'more than 220 genera with about 4.000 species', Neal (1965) thinks 'about 280 genera and 7.000 species' whereas the RHS's Dictionary (by Chittenden & al., 1965) gives 'over 210 genera with about 4.500 species'. Evidently the differences are considerable, and we rely here on the latest indication given in the 'Willis Dictionary': 300 genera and 5.000 species; and this after the separation of the Androstachydaceae, Bischofiaceae, Hymenocardiaceae, Peraceae, Stilaginaceae, Uapacaceae, and most components of the Pandaceae.

To this family belong a number of very important commercial species, some highly poisonous plants, many greatly prized ornamentals and more than enough cosmopolitan weeds. There are trees and shrubs, vines, herbs and succulents. I believe the most complex group is the genus *Euphorbia* itself which many taxonomists have tried to split into more 'reasonable' micro-genera, rejected in turn by other specialists. From the physiognomists point of view this interesting genus of the Old World has developed many of the characteristics of the cacti in the so-called New World. And as one can't say, just like this, which 'World' is older, who can say which family 'copied' from the other; in this particular case similar environmental conditions produce similar growth forms. On the other hand, zoologists for example tend to laugh when told that a common garden weed known as *Euphorbia peplus* belongs to the same genus as the tree-shaped *E.candelabrum*; but they may laugh because they are zoo-logists!

To the Euphorbiaceae belong: *Hevea brasiliensis* (H.B.K.) Muell.-Arg., the Para Rubber-tree (occasionally tried in these islands), and some fine ornamental shrubs such as the widely cultivated *Acalypha* species, the many forms of the Croton complex (*Codiaeum variegatum* (L.) Blume), the Snow Bush (*Breynia nivosa* (Bull.) Small), *Phyllanthus* species and *Jatropha*. *Ricinus communis* L. is cultivated but tends to become weedy, and *Manihot esculenta* Crantz provides a staple food for millions in most tropical countries. There are about 70 species of Euphorbiaceae in the Canary Islands; 17 of these are probably native, out of which 8 are endemics including the famous *Euphorbia canariensis*. Hybridization is not uncommon.

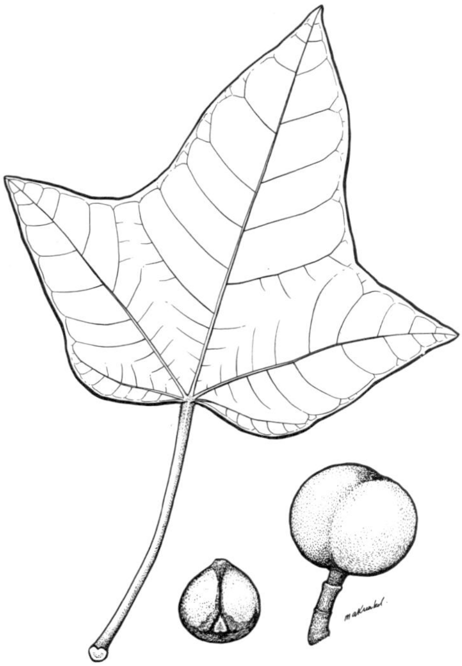

Fig. 59. *Aleurites moluccana*; leaf and fruits = 2/5

Aleurites moluccana (L.) Willd. – Candlenut Tree

This quite decorative tree is native in tropical Asia and some Pacific archipelagos. Usually up to 15 m tall, with an open spreading crown. Trunk short, 40 to 60 cm in diameter; bark grey and furrowed. Briefly deciduous. Leaves on terminal branchlets, alternate, long-stalked. Blade almost herbaceous, shiny green above and tomentose beneath, 3 to 5-lobed, up to 20 cm long and wide. Flowers small, whitish, in terminal racemes. Fruits 1 to 3-seeded, nut-like but less hard-shelled, up to 4 cm in diameter. Propagated from seeds and cuttings, the species grows best at lower altitudes. Sometimes cited as *Aleurites triloba* Forst. The nuts produce the Lumbang or Candlenut Oil used for varnish, paints, soap production, candle-wax as well as for several colourants. The wood is not durable and of little value. Seeds are edible but they should be roasted, a fact not known to the present author; however he survived having read only afterwards about the poison in raw kernels. – For further information on use and properties of this species see also Burkill (pp. 92–96). Alternative names: Varnish Tree, Candleberry Tree, and Indian Walnut.

Ref.: Adams; Barrett 56; Bircher; Burkill; Chittenden; Corner; Encke; Kunkel 69; Little & al.; Menninger 64; Neal; (Uphof).

Hura crepitans Linné – Sandbox Tree

A fine small ornamental tree up to 10 m tall, with a pronounced trunk 40 to 60 cm in diameter; bark grey, rough or scaly. Crown open and spreading; branches with leaf-scars and occasionally with thorns on younger branches. Deciduous. Leaves alternate, stalked, in terminal position; blade ovate to heart-shaped, herbaceous, up to 12 cm long and 7 cm wide; margin somewhat serrate, apex pointed; very conspicuous lateral nerves. Flowers purplish, small; male flowers crowded in cylindrical spikes, the female ones usually single. Fruit-capsule subglobose, adpressed at both extremes; woody, dark grey to brownish, up to 8 cm in diameter, strongly ribbed. Seeds disk-like, greyish-brown.

According to Burkill the fruits 'explode violently when ripe, and from the noise they make, arises the West Indian name for the tree: The Monkey's Dinner Bell'. The noise of such an explosion can be confirmed as we had a fruit, varnished, for almost ten years until one evening we got our fright – and had to collect seeds and the remains of the exploded capsule from every corner of our library. Barrett states that 'this is quite a dangerous tree: it stabs, poisons, and shoots'. Besides this peculiar behaviour: almost all parts of the tree are purgative or poisonous, the latex produces a fish-poison, but the wood is strong and makes a good timber for carpentry and small constructions. The species is sometimes seen as a roadside tree, is propagated from seeds and rooted cuttings, and is native in tropical America.

Ref.: Adams; Barrett 56; Burkill; Chittenden; Graf; Huxley; Irvine; Keay & al.; Kunkel 69; Little & Wadsworth; Long & Lakela; Menninger 67; Neal; Uphof; Webster & Burch.

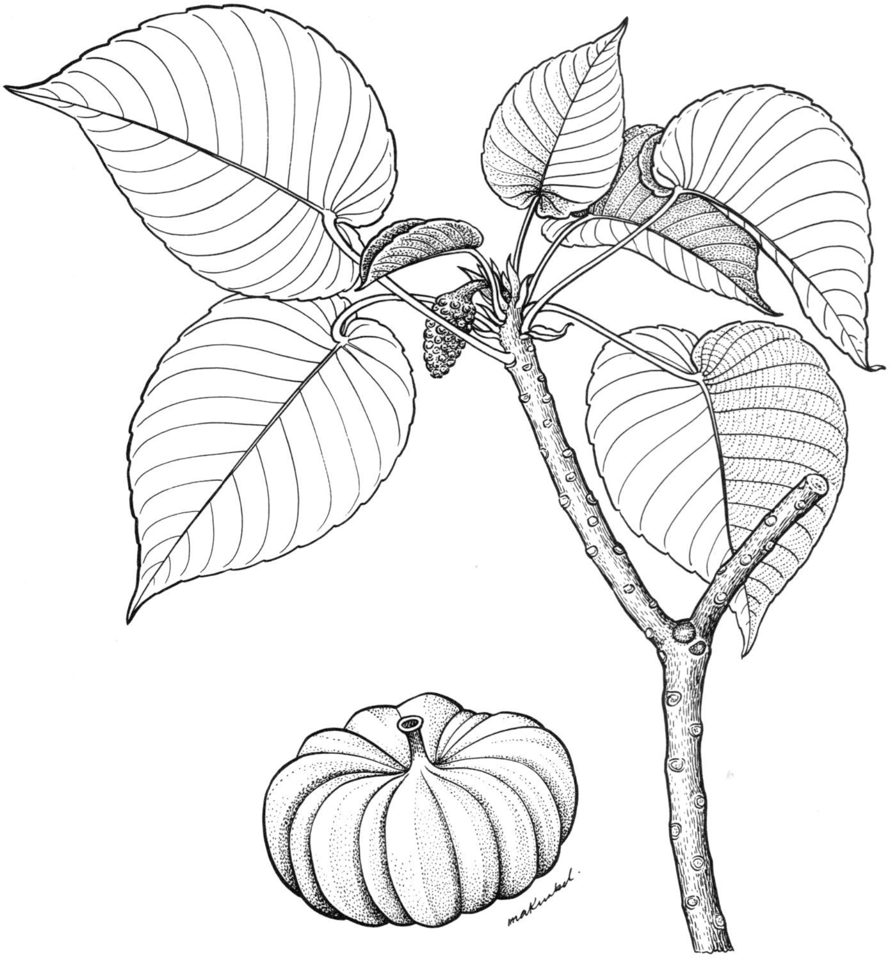

Fig. 60. *Hura crepitans*; flowering branch and fruit = 3/5

PITTOSPORACEAE, the Pittosporum family

A family of 9 genera and about 200 species of trees, shrubs and a few climbers, mainly natives of Australia. Of special importance for gardens only the genus *Pittosporum* which consists of about 150 species.

Pittosporum phillyraeoides DC. – Weeping Pittosporum

A shrubby tree up to 10 m tall, with spreading and pendent branches. Trunk short and slender. Leaves alternate, linear, short-stalked, pale green, up to 10 cm long and, usually, only 0,5 cm wide; subcoriaceous, apiculate or mucronate. Flowers solitary, stalked, cream-coloured, in leaf-axils; rarely in groups. Fruit globose, more or less oval-shaped, yellowish; seeds single, orange-red.

This evergreen species is a native of Australia. Uphof calls the tree 'Butter bush', and says that its seeds are bitter but are ground into a flour which is consumed by the aborigenes. This species is rather rare in our gardens and should be more frequently planted. – I do have to disagree with the late Mrs. Neal that 'a species formerly determined as *P.phillyraeoides* DC., from Australia, has been redetermined as *P.undulata*'; this may refer to 'P.phillyraeoides' of Haworth, not of DC.

Ref.: Chittenden; Encke; Marthi; Menninger 64; (Neal); Uphof.

Fig. 61. *Pittosporum phillyraeoides*; all drawings = 3/5

Pittosporum tobira (Thunb.) Aiton – Japanese Pittosporum

This is a smaller tree reaching only 4 to 6 m height; sometimes of shrubby appearance or planted to form an effective hedge. Trunk slender, often twisted, up to 20 cm in diameter; bark dark grey and somewhat scaly. Crown dense and flattened. This species is also evergreen. Leaves alternate but crowded, short-stalked, oblong-obovate, leathery, dark green above and paler beneath; up to 10 cm long and 3 cm wide. Margin entire but curled down and inwards. The leaves have quite a definite odor. Flowers fragrant, white turning yellowish, in dense terminal clusters. Capsule ovoid or almost triangular, up to 1,5 cm long and finely pubescent; style persistant. Seeds small, reddish and rather sticky.

Native in East Asia (China, Japan), this species was once much cultivated in these islands. However it was such a favourite for attack by scale insects, aphids, and a black fungus, that it fell 'out of fashion'. Planted near to the sea much of this unpleasant trouble is avoided. – There is also a variegated form: cv. 'Variegata' hort.

Ref.: Bircher; Chanes; Chittenden; Eliovson; Encke; Franco 64; Graf; Kunkel 69; Lombardo 61; Marthi; Menninger 64; Moeller 71; Neal; Polunin & Smythies.

Fig. 62. *Pittosporum tobira*; flowering branch and fruit = natural size

Pittosporum undulatum Vent. – Mock Orange

Usually a small tree 6 to 8 m in height; Adams records trees up to 10 meters, whereas Marthi gives trees and shrubs 6 to 15 m tall. The trunk is slender and relatively short; our specimens hardly exceed 25 cm in diameter at chest-height. Bark dark grey and finely fissured; branches verticillate, with lenticels. Crown dense and spreading. Foliage evergreen. Leaves stalked, in whorls or nearly so; blade more or less lanceolate, subcoriaceous, dark green, and delicately veined; up to 15 cm long and 3,5 cm wide, with a softly undulate margin. Flowers whitish, very fragrant, in leafy terminal clusters. Fruits also showy while ripening: orange-coloured globose capsules with a persistant style. The seeds are reddish or purple, and are 'covered with glutinous adhesive' (Adams).

The Mock Orange is native in Australia (a variegated form is also known), and is grown in many parks and gardens, especially of more subtropical zones. It is propagated from seeds and from cuttings, and easily becomes naturalized as birds distribute the sticky seeds; we have seen shrubby specimens up to 2 m tall growing as epiphytes on palm trees (*Phoenix canariensis*). According to Menninger the species 'is so common in Australia and California that it is regarded almost as a weed'; the same author calls the tree the 'Victorian Box', and says it is resistant to salt spray and therefore much used in seaside gardens. The species is well enough known in the Mediterranean area to be included in 'Flora Europaea' (Franco). – Additional names: Native Daphne, and Victorian Laurel.

Ref.: Adams; Bircher; Chittenden; Encke; Franco 64; Graf; Harrison; Kunkel 69; Lombardo 58; Marthi; Menninger 62, 64; Moeller 71; Polunin & Smythies.

A fourth species – *Pittosporum crassifolium* Sol. ex Putterl. – from New Zealand is most likely to occur in these islands. More recent research in gardens revealed also the existance of a few specimens of *Hymenosporum flavum* (Hook.f.) F.v.Muell., a tree with obovate-oblanceolate leaves up to 20 cm long, and with larger, yellowish flowers; its capsule (2,5 cm long) contains winged seeds. The species is native in Australia and New Guinea.

Fig. 63. *Pittosporum undulatum*; all drawings = 1/2

ROSACEAE, the Rose family

Even after the segregation of the Chrysobalanaceae (10 gen., 400 spp.) and Neuradaceae (3 gen., 10 spp.) still a large and rather heterogenous family of trees, shrubs, herbs and some semi-climbers, from every climatic region of the world. About 100–110 genera with perhaps more than 2.000 species many of which are very important fruit trees or shrubs, or are prized as ornamentals (*Cotoneaster*, *Pyracantha*, *Rosa*, *Spiraea* etc.). – In the Canary Islands about 30 species of which 10 or 12 (all more or less woody) are considered as endemics; *Laurocerasus lusitanica* ssp. *hixa*, *Bencomia caudata* and *Marcetella moquiniana* are of tree-like habit.

Amygdalus communis Linné – Almond Tree

The well-known, widely cultivated Almond tree is thought to be native in the Middle East; frequently also cited as *Prunus communis* Arc., and *P.amygdalus* Batsch. Tree 6 to 10 m tall; trunk short, generally irregular (bent or twisted), bark dark grey to blackish in older specimens, and deeply furrowed. Crown usually spreading. Foliage deciduous. Leaves stalked, alternate but clustered, often leaving somewhat warty protuberances; blade oblong-lanceolate, herbaceous to subcoriaceous; size and shape depending on the particular variety of which very many are known. Upper surface usually rather shiny. Flowers solitary or in pairs, very showy, with a white or pinkish corolla up to 5 cm in diameter. Fruit (a drupe) velvety, thin-fleshy at first becoming hard-skinned, and enclosing a single 'nut' with a woody indehiscent shell 3 to 4 cm long, pale grey or greyish-brown. The seed itself is the well-known almond which is not only edible but highly appreciated.

An undemanding tree suitable for plantation in a more arid environment. Propagation from seeds and basal shoots. Good varieties are obtained by grafting only, and the stock of the Almond Tree serves also for grafting related species (i.e. Peach and Apricot). There are two main varieties: the Sweet and the Bitter Almond both of which have their uses. Well cared for trees may produce up to 20 kg of almonds (with shells), per year. The wood is hard and used for making small items such as bowls, plates, forks, spoons etc.; also as fuel. The woody shells also make a good combustible and are, besides this, used for 'maturing' wines and brandies.

The Almond tree has been cultivated since antiquity, and is now extensively grown around the Mediterranean basin and in California where numerous new varieties have been created (Kèster, López). Almonds are eaten raw, roasted, salted and in pastries; ground almonds are the basis of marzipan and used in confectionary. The oil obtained from the seed is non-drying and applied as emollient, demulcent, sedative, for coughs, etc. (Uphof). Almond trees are also fine ornamentals, and a 'flowering hillside' makes an unforgettable impression.

Ref.: (Beyron); (Bircher); (Chanes); (Eliovson); (Encke); (Fitschen); (Font Quer); (Harrison); (Kester); (Lombardo 58); López; (Moeller 71); (Perry); (Polunin & Huxley); Rauh; Ruiz de la Torre; Uphof.

Related to the Almond tree are

Amygdalus persica Linné – Peach Tree

Synonym: *Prunus persica* (L.) Batsch. Another small tree, native in China, with very similar leaves but large, tomentose, juicy and edible fruits up to 9 cm in diameter. The seeds are similar to those of almonds but are bitter and have a slight cyanide taste; eaten in quantities may cause a serious intoxication. The species hybridizes with the Almond (= *A.* × *amygdalopersica*). And

Armeniaca vulgaris Lam. – Apricot

A small tree also probably from China, with deciduous, long-stalked, pointed, rather herbaceous elliptic leaves and soft tomentose, yellowish or red-spotted fruits. The seeds are sometimes mixed with or used as a substitute for the true almond. The species is frequently cited as *Prunus armeniaca* L.

Somewhat related but 'once removed' are the Cherries, with *Prunus avium* L. (Sweet Cherry) and *P.cerasus* L. (Sour Cherry), also the Plums, with *P.insititia* L. (Damson Plum, often cited as a subspecies of the following), *P.domestica* (the common Garden Plum which may be of hybrid origin), and others. There are several species of purely ornamental Cherries and Plums, none of them frequent in these gardens. Remotely related to the above cited group are the Apples (i.e. *Malus pumila* Mill.) and Pears (*Pyrus communis* L.), all of which are rich in varieties and cultivars.

Fig. 64. *Amygdalus communis*; all drawings = 3/5

Eriobotrya japonica (Thunb.) Lindley – Loquat Tree

The Loquat, Japanese Plum or Japanese Medlar, although its name indicates Japan as its home country, is thought to be native in China. However it is extensively cultivated in Japan (over 800 varieties!) and elsewhere in subtropical countries; not suitable for the humid tropics.

Small tree 4 to 7 (10) m tall; trunk short, 12–20 cm in diameter. Bark dark grey, fissured or scaly. Crown rounded or spreading; branches easily breaking off, twigs somewhat villous. Foliage evergreen. Leaves alternate, short-stalked, stiff-leathery, coarsely toothed, up to 25 cm long and 7 cm wide; blade oblong, dark green and glossy above, rusty-villous beneath; venation deep set so making a ridged surface. Young leaves pale reddish and woolly all over. Flowers white, fragrant, in stout terminal inflorescences 8 to 12 cm long. Fruits spherical, ellipsoid or pear-shaped (varieties!), from off-white, yellow, golden yellow to brownish-orange, soft-skinned, juicy, up to 4 cm in diameter; edible. Seeds 1 to 3(5), shiny dark brown.

Propagated from seeds and best grown in somewhat cooler climates; however quite drought-resistant, and recommended for seaside gardens by Menninger. Grafting is also practiced. Fruits are eaten raw or in preserves and jellies; slightly laxative. According to Kennard & Winters 'the tree is subject to attacks by fire-blight, a bacterial disease' which is best controlled by removing affected parts and applying very little nitrogen.

Ref.: Ball 68; Barrett 56; Bircher; Burkill; Calabria; Chittenden; Encke; Graf; Harrison; Kennard & Winters; Kunkel 69; Menninger 64; Molesworth Allen; Mortensen & Bullard; Neal; Perry; Polunin; Polunin & Huxley; Polunin & Smythies; Schaeffer; Uphof.

Fig. 65. *Eriobotrya japonica*; drawings = 1/2

Laurocerasus lusitanica (L.) Roemer – Portugal Laurel

Usually cited as *Prunus lusitanica* L., this more precisely 'Portuguese Laurel Cherry' seems to be the typical subspecies whereas ssp. *hixa* is native in these islands (Kunkel & Kunkel). An evergreen tree 10 to 15 m tall; trunk short, bark grey and rather smooth. Crown more or less pyramidal, lower branches spreading or pendent and quite flexible. Petioles and young twigs red. Leaves alternate, stalked, oblong-lanceolate to ovate, dentate-serrate, subcoriaceous, dark green, up to 15 cm long and 5 cm wide; young leaves somewhat folded. Flowers small, white, very fragrant, in showy often erect axillary racemes 15 to 25 cm long. Fruits ovoid but pointed, little fleshy; dark purple or almost black, 8 to 12 mm long; not edible.

This ornamental tree is native in the Iberian Peninsula and is planted in cooler climates where it withstands an occasional frost. Propagation from seeds and basal shoots. – The related *Laurocerasus officinalis* Roem., the true Cherry Laurel, is very rare in Canary gardens; its leaves are wider, almost entire, and the inflorescences are shorter and more crowded.

Ref.: (Chittenden); (Eliovson); (Encke); (Fitschen); (Perry); (Polunin); (Polunin & Smythies); (Webb 68).

Fig. 66. *Laurocerasus lusitanica*; flowering branch = 3/5, solitary flower = natural size

LEGUMINOSAE

Although Kew Herbarium and some other well-known institutions treat the Leguminosae as a complex family, I confess belonging to the 'splitters' preferring more numerous but smaller groups, accepting the Leguminosae as an *order* divided into three distinctive families as dealt with below.

MIMOSACEAE, the Mimosa family

The Mimosa family s.str. consists of about 40 genera with some 2.200 to 2.500 species. The largest genera are *Acacia* (700–800 spp.) and *Mimosa* (450–500 spp.) followed by *Inga*, *Pithecellobium*, *Albizia*, *Calliandra* etc. Some species are important timber trees, others are appreciated for reafforestation or may have medicinal properties; most are natives of drier countrysides. Some *Acacia* and *Albizia* rapidly naturalize when introduced into suitable environments. There are no native Mimosaceae in the Canary Islands.

This account may well begin with *Acacia*, a genus of evergreen or deciduous species most abundant in Australia. However *Acacia* trees and shrubs are also of phytosociological importance in landscapes of the Middle East, North and South Africa, and at lower altitudes in drier parts of North and South America including the West Indies. In its natural distribution the genus never reached Europe; however a number of species have been introduced and have proved very successful in parks and gardens. Most Acacias grown here are large shrubs or small trees but *Acacia baileyana*, *dealbata*, *decurrens* and *melanoxylon* develop into trees of considerable size. All Acacias are propagated from seeds; some of them should be previously treated (hot water, fire, etc.).

Group of species with reduced leaves or phyllodes:

Acacia melanoxylon R.Br. – Australian Blackwood

The largest of all species grown in these islands. Tall tree up to 30 meters in height, with an almost pyramidal, dense and rather dark crown. Trunk straight, 50 to 70 cm in diameter; bark dark grey and fissured. Leaves mainly reduced to phyllodes but true bipinnate leaves are also produced, especially on younger branches. Young shoots somewhat angular and downy. Phyllodes lanceolate or oblanceolate, falcate, up to 12 cm long and 2,5 cm wide, dark green, subcoriaceous, with several prominent parallel veins. Flower-heads globose, yellow, about 6 mm in diameter, several together forming short axillary racemes. Pods flattened, narrow, slightly constricted, up to 10 cm long, usually curved; seeds few, bedded in a red soft funicle.

Probably native in Tasmania and Southern Australia. Wood hard, used for furniture, cabinet-making, boat-building, tool handles, billiard-tables etc. (Uphof). A good shade and roadside tree, often planted on eroded slopes but easily becoming 'weedy'. Hardy in parts of Britain and S.W. Europe.

Ref.: Bean; Bircher; Chanes; Chittenden; Encke; Franco 68; Harrison; Lombardo 58; Menninger 64; Neal; Polunin & Smythies; Uphof.

Fig. 67. *Acacia melanoxylon*; flowering branch and fruit = 3/5, solitary seed = × 1,5

Acacia longifolia (Andr.) Willd. – Sydney Golden Wattle

A small tree native in Southern Australia, up to 10 m in height. Crown open, with spreading branches. Trunk slender, often twisted; bark grey and smooth. Young shoots angular. Phyllodes lanceolate or oblong-lanceolate, coriaceous, dark green, 7 to 15 cm long; with numerous parallel veins.

The typical form of this species (var. *longifolia*) seems to be rare in these islands; more frequently cultivated is var. *sophorae* (Labill.) F.v.Mueller which is a large shrub of spreading habit and shorter phyllodes (up to 8 cm long). Flowers very fragrant, small, bright yellow, in axillary cylindrical spikes 3 to 5 cm long. Pods brownish, nearly cylindrical, slightly constricted, 8 to 10 cm long; 6 to 10-seeded, funicle short.

The Sydney Golden Wattle makes a good ornamental for gardens. Franco reports that the species is planted for stabilizing coastal dunes. If Uphof's 'A.cibaria' is identical with this species, then seeds are said to be eaten by the aborigenes of Australia.

Ref.: Bean; Bircher; Chittenden; Eliovson; Encke; Franco 68; Graf; Harrison; Lombardo 58; Menninger 64(?); Polunin & Smythies; (Uphof ?).

Fig. 68. *Acacia longifolia* var. *sophorae*; flowering branch and fruits = 2/3, seed = × 1,5, solitary flower = × 3

Acacia cyanophylla Lindley – Blue-leaved Wattle

Native in Western Australia and well established in the Middle East, parts of Southern Europe and the Canary Islands. Often suggested for reafforestation in arid zones by Good & Barney, Kaul & al., and other authors.

Small tree up to 7 m in height; branches upright, later arching, pronouncedly zigzagged. Trunk short; bark dark grey and fissured. Phyllodes glaucous, coriaceous, broad lanceolate and strongly undulate, 12 to 15 cm long and 1,5 to 2,5 cm wide; strong central vein with fine, hardly visible lateral veins. Flowers deep to golden yellow, in spherical heads up to 1 cm in diameter, somewhat spaced in axillary racemes up to 8 cm long (which might become leafy). Pods narrow, brownish, almost straight, flattened, up to 10 cm long and somewhat constricted; 8 to 15-seeded, funicle small and pale.

Easily grown from seeds of which, according to Goor & Barney there are about 60.000 per kg. A fast growing species originally much planted but later replaced by the following, because the Blue-leaved Wattle is often infested by scale and other insects.

Ref.: Bircher; Chittenden; Franco 68; Goor & Barney; Kaul & al.; Kunkel 69; Lombardo 58; Moeller 68; Polunin & Smythies; Schaeffer.

Fig. 69. *Acacia cyanophylla*; fruiting branch and flower heads = 3/5

Acacia saligna (Labill.) Wendl. – Australian Weeping Wattle

This closely related species also from Western Australia differs from the above by its greener, narrower and straight phyllodes, smaller and paler but more prolific flower-heads, and narrower pods. Sometimes it is difficult to distinguish the Weeping from the Blue-leaved Wattle, and it is very possible that hybridization occurs. – Uphof cites *A.saligna* as a synonym of *A.decurrens*, an opinion very difficult to understand.

Small tree, 4 to 8 meters tall; openly branched, crown spreading. Phyllodes spaced, more or less linear, 12 to 20 cm long and usually less than 1 cm wide; subcoriaceous, greenish, almost straight. Flowers pale yellow, heads less than 1 cm in diameter, on slender, occasionally leafy panicles up to 10 cm long. Pods up to 15 cm long, flattened, constricted, with up to 15 seeds each.

Ref.: Bircher; (Beyron ?); Chittenden; Encke; Graf; Harrison; Kunkel 69; Moeller 71; (Uphof ?).

Other related species such as *Acacia aneura*, *A.cyclops* and *A.retinodes* are more bushy and will be dealt with in a forthcoming volume.

Fig. 70. *Acacia saligna*; drawings = 3/5

Photo 2. Wintery aspect of an old fig tree (*Ficus carica*)

Photo 3. 'Manipulated landscape (Canary Islands), with *Eucalyptus* trees.

Species with bipinnate leaves:

Acacia decurrens (Wendl.) Willd. – Green Wattle

A very fast growing tree up to 20 m tall, native in Australia where, according to Burkill, it may reach up to 50 m in height; however in his otherwise excellent account *A.decurrens* and *A.dealbata* are not separated, and his data may refer to the following species described as Silver Wattle.

The proper Green Wattle has a pronounced trunk 40 to 60 cm in diameter; bark brownish, furrowed or scaly. Branches upright, the younger ones slender and spreading. Leaves up to 12 cm long, deep green and only minutely hairy. Each leaf consists of 13 to 15 pairs of pinnae, with up to 50 pairs of tiny leaflets each. The leaves are stalked, and there are pronounced glands on the rachis. Flowers pale yellow, fragrant, in globose heads up to 8 mm in diameter forming a showy panicle 10 to 12 cm long. Pods flattened, 6 to 10 cm long, dark brown, constricted; 8 to 12-seeded.

This tree is also sometimes called Black Wattle, or Greenwattle Acacia. It grows easily from seeds reaching 4 to 5 m in height in less than 5 years. It is an excellent species for planting in semi-arid ground, and is planted for fuel in East Africa. Uphof reports that the bark is used for tanning.

Ref.: Bircher; (Burkill ?); Chittenden; Encke; Harrison; Moeller 68; Neal; Rauh; (Uphof ?).

Fig. 71. *Acacia decurrens*; general drawing and solitary (ripe) fruit = 3/5, solitary pinna = × 1,5, leaf detail = × 3

Acacia dealbata Link – Silver Wattle

Also of Australian origin, this very successful tree is now widely distributed and cultivated in many countries. Small potplants are often sold as 'Mimosas'. Flowering branches from trees grown in the Mediterranean region are sent to many European flowershops. – The species is closely related to that previously described and is treated by several authors as a variety. Uphof includes it as a synonym in *A.decurrens* but describes it some lines above (p. 4) as a separate species.

Tree up to 30 or 40 m in height. Trunk 50 to 80 cm in diameter; bark relatively thin, greyish, almost smooth. Crown large and rather dense; twigs somewhat downy. Leaves up to 10 cm long, with 8 to 12 pinnae each of which bears about 30 pairs of very small leaflets 4 to 6 mm long, of a silvery-glaucous colour and finely downy. Glands on rachis small. Flowers pale yellow to somewhat greenish, in spherical heads forming rich-flowering panicles up to 12 cm long. Pods rather flat, little constricted or not at all, brownish-blue, 9 to 12 cm long and slightly hairy. Seeds 8 to 12.

Although this species makes a fine solitary or roadside tree, in favourable climates it easily becomes 'weedy' and hard to control; it has proved aggressive in Northwestern Spain, Madeira, in Southern Chile and other places. – According to Uphof the Silver Wattle produces 'a gum exceedingly viscose and as useful as Gum Arabic'.

Ref.: Bean; (Burkill); Chanes; Chittenden; Encke; Franco 68; (Graf); Harrison; (Lombardo 58); Mitchell; (Neal); Polunin & Smythies; Uphof (?).

Acacia baileyana F.v.Mueller – Cootamundra Wattle

A smaller tree from New South Wales (Australia), of more slender habit with often pendulous branches. Branches, leaves and pods glabrous. Leaves glaucous, up to 6 cm long and very broad when compared with the Silver Wattle; pinnae 4 to 6, with pronounced glands on the rachis; each pinnae with 15 to 25 pairs of leaflets 6 to 8 mm long. Flower-heads golden yellow. Pods 8 to 12 cm long, 8 to 10-seeded.

The Cootamundra or Bailey Wattle, according to Menninger (62) 'is so common and pops up everywhere from seed, growing quickly and standing up to a long dry hot summer, that we often treat it as little more than a weed, but this is most unjust to a beautiful Wattle. The massed blossom on *baileyana* in early winter is a glorious sight. It is a highly-prized ornamental tree'.

Ref.: Bean; Chittenden; Eliovson; Encke; Graf; Harrison; Lombardo 58; Menninger 62; Neal.

It is interesting that all *Acacia* species from Australia cultivated in these islands are unarmed whereas the African and American species all seem to be spiny.

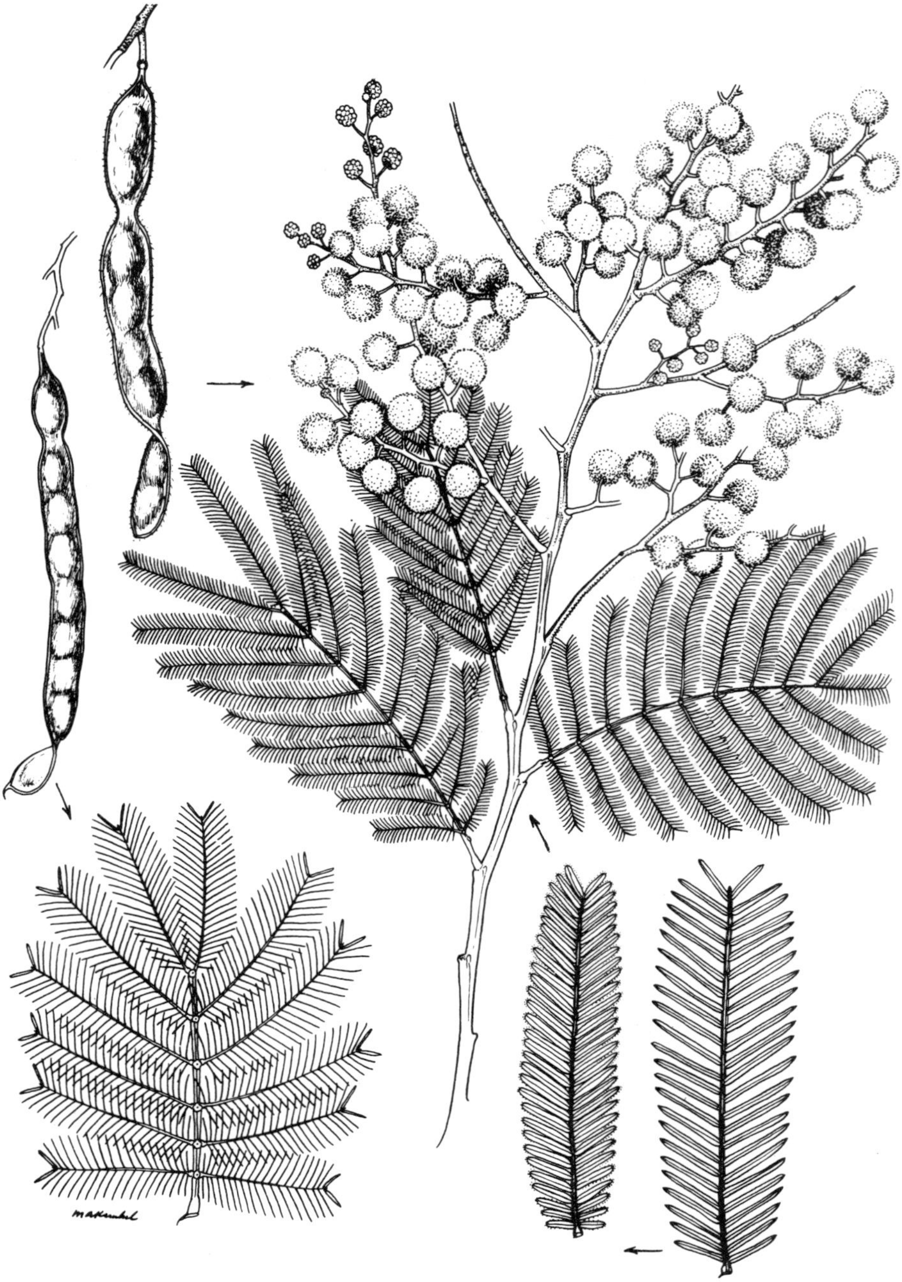

Fig. 72. *Acacia dealbata* (main flowering branch and slightly hairy fruit) and *A.baileyana* (below, left); main drawings = 3/5, leaf details = 1,5

Acacia karroo Hayne – Sweet-thorn

This admirable tree of South African origin is often cited as Allthorn Acacia or Karoo Thorn, or mentioned under its synonym *Acacia horrida* Willd. – And although Franco (1968: 84) cites the species as 'A.karoo' we may insist on the double 'rr' in its name as given by Kunkel, Palmer & Pitman, and others. The last mentioned authors, in their excellent treatise of South African trees dedicate seven pages to this species.
The Sweet-thorn is a noble tree 8 to 12 m in height. Trunk short and somewhat knotty, up to 80 cm in diameter. Bark dark, rough, fissured or furrowed; branches spiny. Crown rather flat and wide-spreading. The species is evergreen in the Canary Islands but seems to be deciduous in its home country. Leaves bipinnate, bright glossy green, with 3 to 4(6) pairs of main pinnae each with 9 to 15 (rarely up to 20) pairs of leaflets about 1 cm long. Glands on rachis prominent. Spines whitish, rather short in our specimens but apparently over 3 cm long in South African material. Flowers golden-yellow, in globose heads up to 1,5 cm in diameter, stalked and clustered to 4–8 in leaf-axils. Pods narrow, flattened, dark grey or brownish, up to 15 cm long, usually curved and somewhat constricted; 6 to 10-seeded.
Plants easily grown from seeds but of rather slow development. Young plants spiny all over. An excellent species for hedges and for reafforestation of more arid land. If well formed it makes a good shade tree. According to Palmer & Pitman its flowers provide a good forage for bees; monkeys eat the pods, leaves are consumed by stock, branches are used to make kraals and to protect plant nurseries etc.; the wood is employed for fuel and building purposes, the bark for tanning, for cord and mat making, and also has medicinal properties. Larger spines are used as pegs or pins. According to the same authors fine-shaped solitary specimens 'often with yellow blossom makes many a Karoo Christmas tree'.

Ref.: Eliovson; Franco 68; Kunkel 69; Menninger 62; Palmer & Pitman; Polunin & Smithies; (Uphof); Van der Spuy.

Other South African Acacias, more recently introduced to the Canary Islands (and not illustrated in this account), include:

Acacia giraffae Willd. – Camel-thorn

Tree with a twisted trunk; leaves with 3 to 4 pairs of pinnae and 10 to 15 (18) pairs of leaflets each. Thorns brownish, very strong and inflated at the base. Pods bulgy and very hard, velvety, up to 12 cm long, 4 to 5 cm wide and about 1,5 cm thick. And

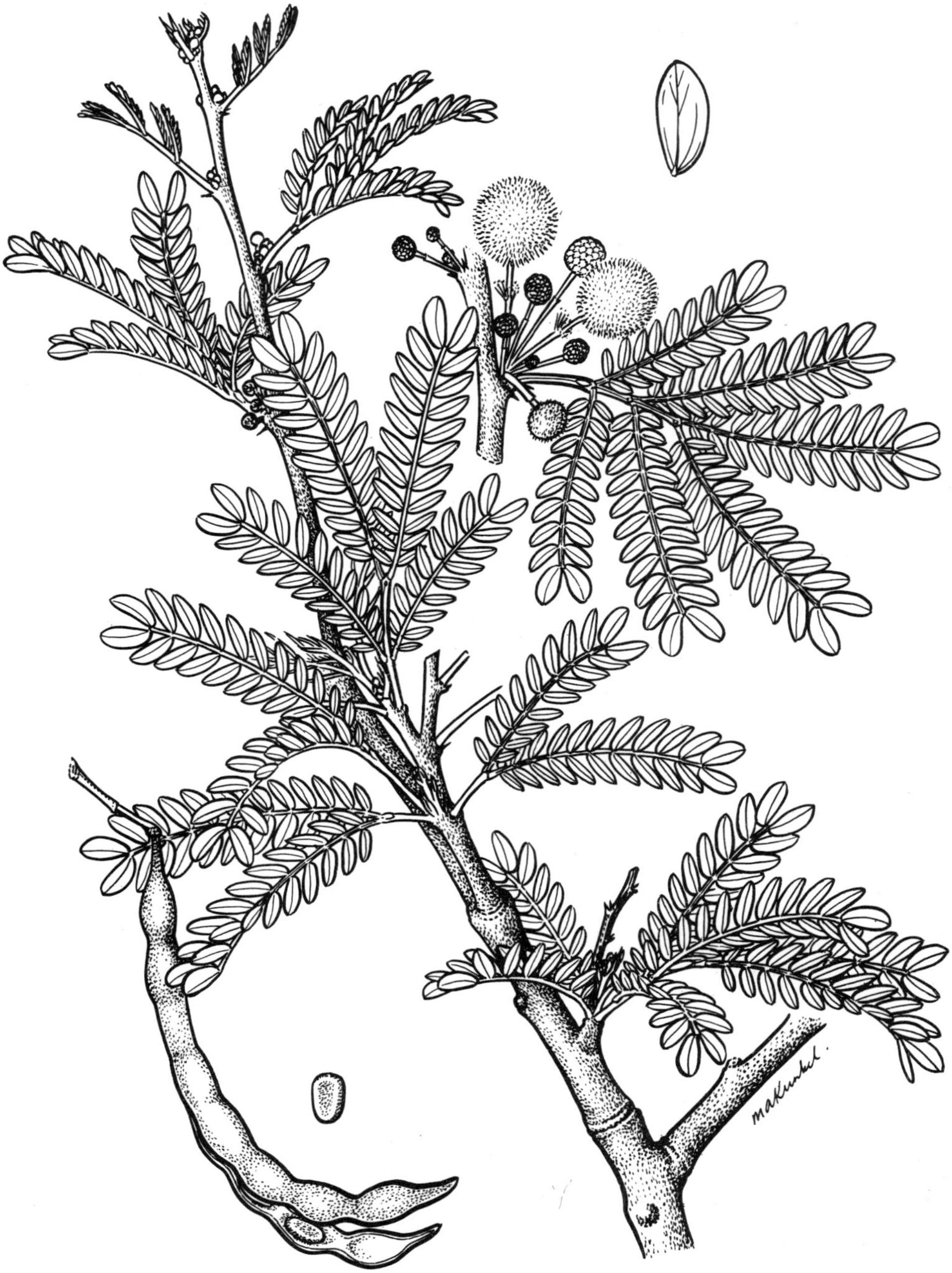

Fig. 73. *Acacia karroo*; flowering branch and fruit = 3/5, solitary leaflet = 1,5

Acacia galpinii Burtt Davy – Monkey-thorn

A larger tree reaching over 20 cm in height, with spreading branches. Leaves with 6 to 8 (10) pairs of pinnae with up to 30 pairs of leaflets. Flowers in clustered spikes, cream-coloured, honey-scented. Pods large and woody, pale brown, 15 to 20 cm long and about 2,5 cm wide.

From North Africa (Western Sahara) seeds were imported of

Acacia raddiana Sawi – Talha

which is probably the northern counterpart of the S.African *A.tortilis* (Forssk.) Hayne, and may deserve the same common name: Umbrella Thorn. Both species are closely related; however *A.raddiana* is kept as a separate species in the 'Flora of West Tropical Africa' (I/2: 500), and no reference is made to this name in Palmer & Pitman's 'Trees of Southern Africa'.
Tree up to 10 m tall. Trunk relatively short or branched at the base, often twisted. Crown flat, 'umbrella'-shaped; branches extremely spiny. Foliage briefly deciduous. Leaves bipinnate, with small glands on the rachis; up to 4 cm long, with 3 to 4 pairs of pinnae and 8 to 12 pairs of small, pale green leaflets each. Spines whitish, up to 6 cm long. Flowers pale yellow, in globose heads about 1 cm in diameter appearing singly or in groups up to 4. Pods narrow, flat, curled or twisted, up to 15 cm long.
Leaves eaten by goats. A good and most effective fence or hedge species; better shape is obtained by pruning basal shoots and lower branches.

Ref.: Keay & al.; Ozenda.

Other well-known North African or Oriental species include *Acacia nilotica* (L.) Willd. ex Del., *A.senegal* (L.) Willd., and *A.catechu* Willd., but none of these seem to be cultivated in these islands. A very common American species – the Opoponax or Sweet Acacia, *A.farnesiana* (L.) Willd. – is shrubby and will be dealt with in a forthcoming volume.

Other mimosoid genera:

Fig. 74. *Acacia raddiana*; flowering branches and fruit = 3/5, solitary leaf = × 1,5

Albizia lophantha (Willd.) Bentham – Albizia

This common small tree 6 to 8 (10) meters tall is a native of Western Australia and is naturalized in Hawaii, Madeira, the Canary Islands, Southeast Africa and perhaps elsewhere. The species is sometimes referred to as *Albizia distachya* Macbr.
Trunk slender; bark dark grey, smooth or somewhat fissured; branches upright, easily damaged by strong winds. Deciduous, or evergreen whilst growing in moist woodlands. Leaves up to 30 cm long, bipinnate, with glands on the rachis; 7 to 14 pairs of pinnae, each of which may have up to 40 pairs of dark green herbaceous leaflets. Leaves 'closing' (folding) in the afternoon or in strong sunshine. Flowers cream-coloured, in dense cylindrical spikes up to 8 cm long. Pods flat, brown, 7 to 10 cm long; with 6 to 10 black seeds.
The species is propagated from seeds and is successfully cultivated in coastal gardens if water is freely available. If planted along roadsides in more humid, elevated zones the *Albizia* tends to become a 'weed' invading neighbouring slopes.

Ref.: Beyron; Bircher; Chittenden; Encke; Kunkel 69; Lombardo 61; Menninger 64; (Moeller 71); Neal; Perry; Schaeffer.

Albizia lebbek (L.) Bentham – East Indian Walnut

A more recent introduction is this fast-growing tree native in tropical Asia. The commercial or common name refers to the walnut-brown heartwood which is often used for furniture. The species is grown as a roadside tree in many tropical and subtropical countries including Central and South America, Hawaii, West Africa, and even in the more arid Near and Middle East. It is a popular shade tree of villages around the Persian Gulf where it is peculiar in so far as the leaves (folded at night) attract flies. The dry, rather persistant and rattling seed-pods are the origin of its second vernacular name: Woman's tongue.
The tree may reach 15 to 20 m in height; trunk up to 60 cm in diameter. Bark grey and rough, becoming fissured, and exuding a gum when cut. Crown spreading; foliage deciduous. Leaves bipinnate, up to 20 cm long; 3 to 4 pairs of pinnae with up to 12 pairs of large herbaceous leaflets. Flowers greenish-white or cream-coloured, in long-stalked clusters or false umbels with very long stamens. Pods somewhat papery, greyish-yellow and rather persistant, flat, up to 20 cm long and 3 cm wide.
The wood is used for construction and for fuel. Leaves are eaten by goats and other animals. The bark contains saponin and is sometimes used as a substitute for soap. Burkill reports the medicinal properties of tannin being

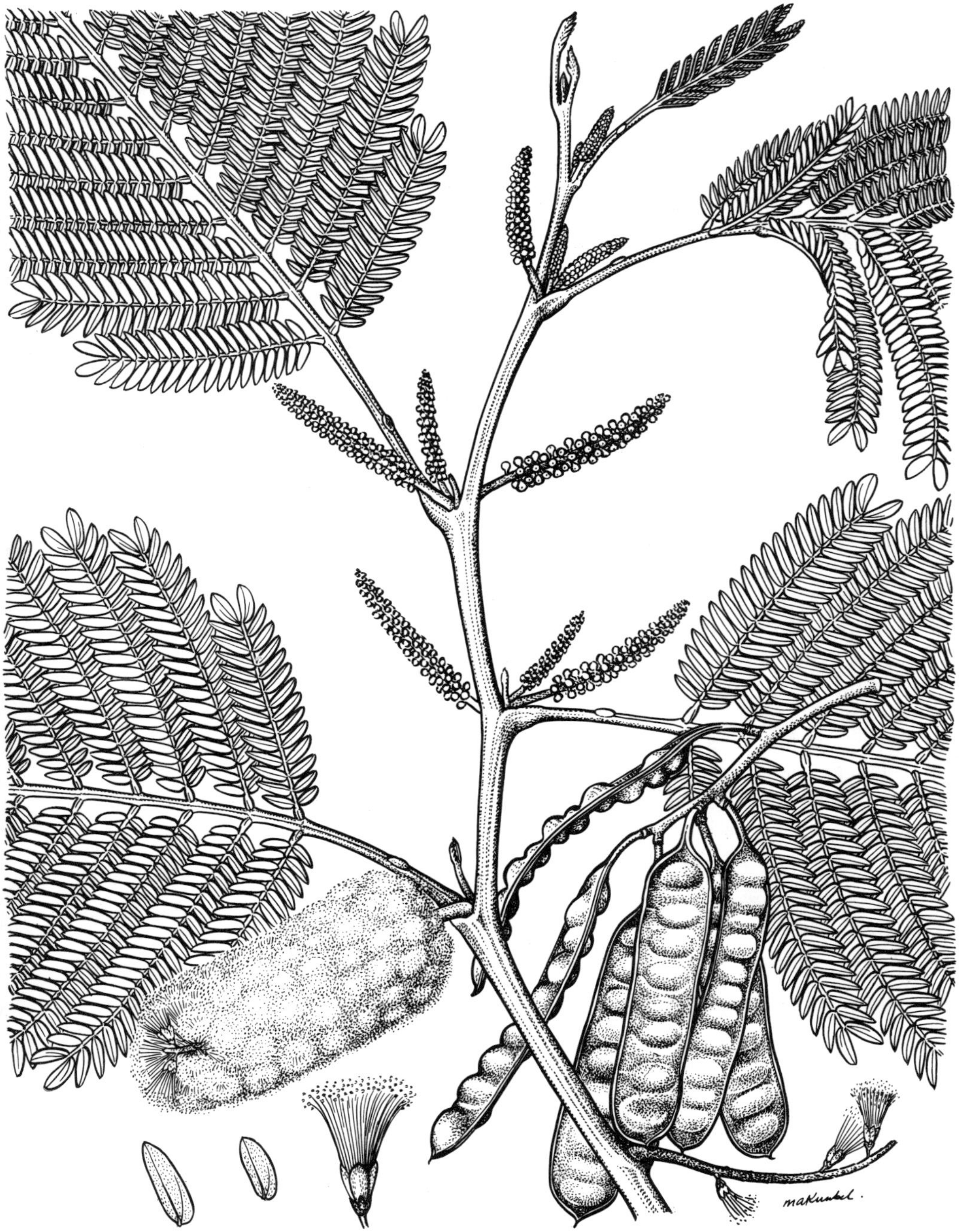

Fig. 75. *Albizia lophantha*; flowering branch and fruits = 3/5, leaflets = natural size, solitary flower = × 1,5

used for dysentery and haemorrhoids. Oil obtained from various parts of the tree is used for treating leprosy in Southeast Asia.

Ref.: Adams; Barrett 56; Bircher; Burkill; Chittenden; Corner; El Hadidi & Boulos; Keay & al.; Little & Wadsworth; Long & Lakela; Menninger 64; Neal; Uphof.

Other species of the genus but hardly ever found in gardens of this region include *Albizia julibrissin* (Willd.) Durazz. (from the Near East), and *A.procera* (Roxb.) Benth. (trop. Asia and Australia). Unfortunately the most attractive African species of this genus are not found in cultivation.

Pithecellobium dulce (Roxb.) Bentham – Monkey Pod

Also known as Guaymochil, Opiuma, Madras Thorn, and Manila Tamarind, this species is native in drier parts of tropical America. A frequently cultivated tree up to 12 m tall, with a wide-spreading crown. Trunk short and soon ramified; bark greyish, fissured, with lenticels. Branches spiny. Briefly deciduous; young leaves usually reddish. Leaves bipinnate although each leaf has only 2 pairs of opposite leaflets; leaflets oblong or obovate, 2 to 4 cm long, emarginate or not. With short spines which are converted stipules, always in pairs. Flowers greenish-white or cream-coloured, in rounded heads, solitary or clustered in more or less terminal position. Pods inflated, curved, pale yellow tinged with pink and mauve, up to 12 cm long, with 5 to 10 black seeds embedded in a whitish pulp.

Propagation from seeds or cuttings. The Monkey Pod makes an effective hedge. The wood is somewhat brittle but used for fuel and simple constructions. The bark contains tannin and a yellow dye. Leaves are eaten by stock. A gum is obtained from the trunk; the pulpy aril surrounding seeds is edible, and a refreshing drink may be prepared from it.

Ref.: Adams; Barrett 56; Bircher; Burkill; Chittenden; Corner; Irvine; Kunkel 69; Little & Wadsworth; Long & Lakela; Martinez; Menninger 64; Neal; Pesman; (Uphof).

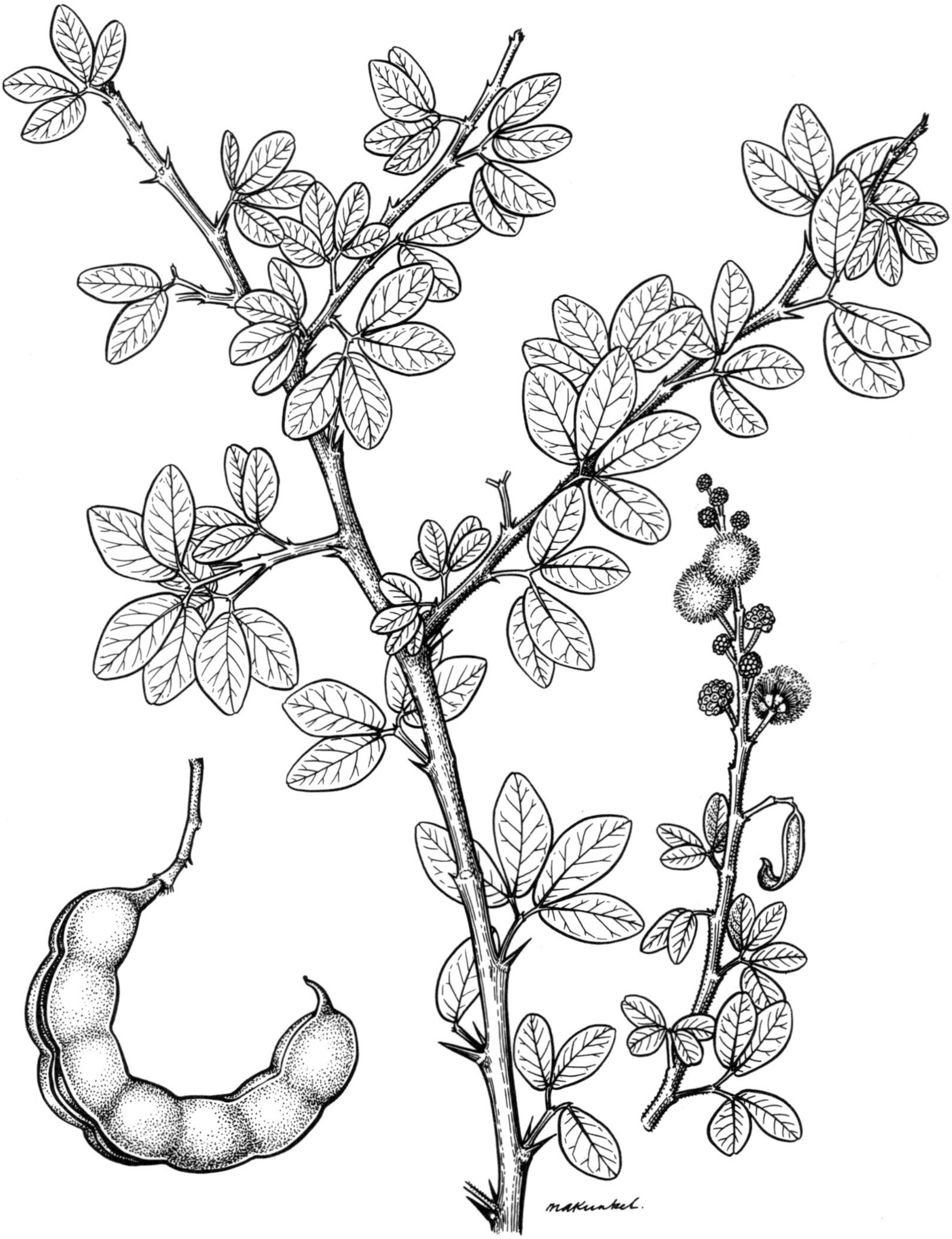

Fig. 76. *Pithecellobium dulce*; all drawings = 3/5

Samanea saman (Jacq.) Merrill – Rain Tree

Often cited as *Pithecellobium saman* (Jacq.) Benth. and cultivated in tropical countries as a shade and roadside tree. Usually 15 to 20 m tall but very old specimens may reach up to 50 m in height. Trunk short, 70 to 100 cm in diameter; bark rough forming corky ridges. Foliage evergreen. Leaves alternate, bipinnate, about 30 cm long with 3 to 5 pairs of pinnae covered by minute hairs; pinnae with 3 to 8 pairs of broad falcate leaflets of which the terminal pair is the largest; glands present. Flowers with long pink stamens in long-stalked open 'heads'. Pods flattened, rather hard, blackish, up to 20 cm long; seeds reddish-brown.

The Rain Tree is native on riversides in drier parts of tropical America. Wood used for fence posts and in rural carpentry. 'The nutritious pods are relished by cattle, hogs, and goats and have a flavour like licorice, which some persons like' (Little & Wadsworth)' The leaves fold up at night or in cloudy weather, a possible indication of approaching rain. – Propagation from seeds and cuttings; a fast growing species for gardens at lower altitudes. In Menninger's 'Fantastic Trees' (1967, p. 217) there is a photograph of a 'gigantic saman tree in Tobago that supported the tree house in Walt Disney's motion picture *Swiss Family Robinson*'.

Ref.: Adams; Barrett 56; Bircher; (Burkill); (Chittenden); (Corner); Graf; Irvine; Keay & al.; (Little & Wadsworth); Menninger 64, 67; Neal; (Uphof).

Also deserving mention but not illustrated:

Prosopis juliflora (Sw.) DC. – Mesquite

Another tree from drier tropical America, and widely naturalized in several places. Deciduous; branches with short spines, leaves with several pinnae and up to 30 pairs of leaflets each. Flowers cream-coloured, very small, in long spikes. Pods flattened; seeds embedded in a sweet edible pulp. And *Enterolobium cyclocarpum* (Jacq.) Griseb., the Elephant's Ear, from South America, as well as species of *Inga* which seem to be popular in American gardens.

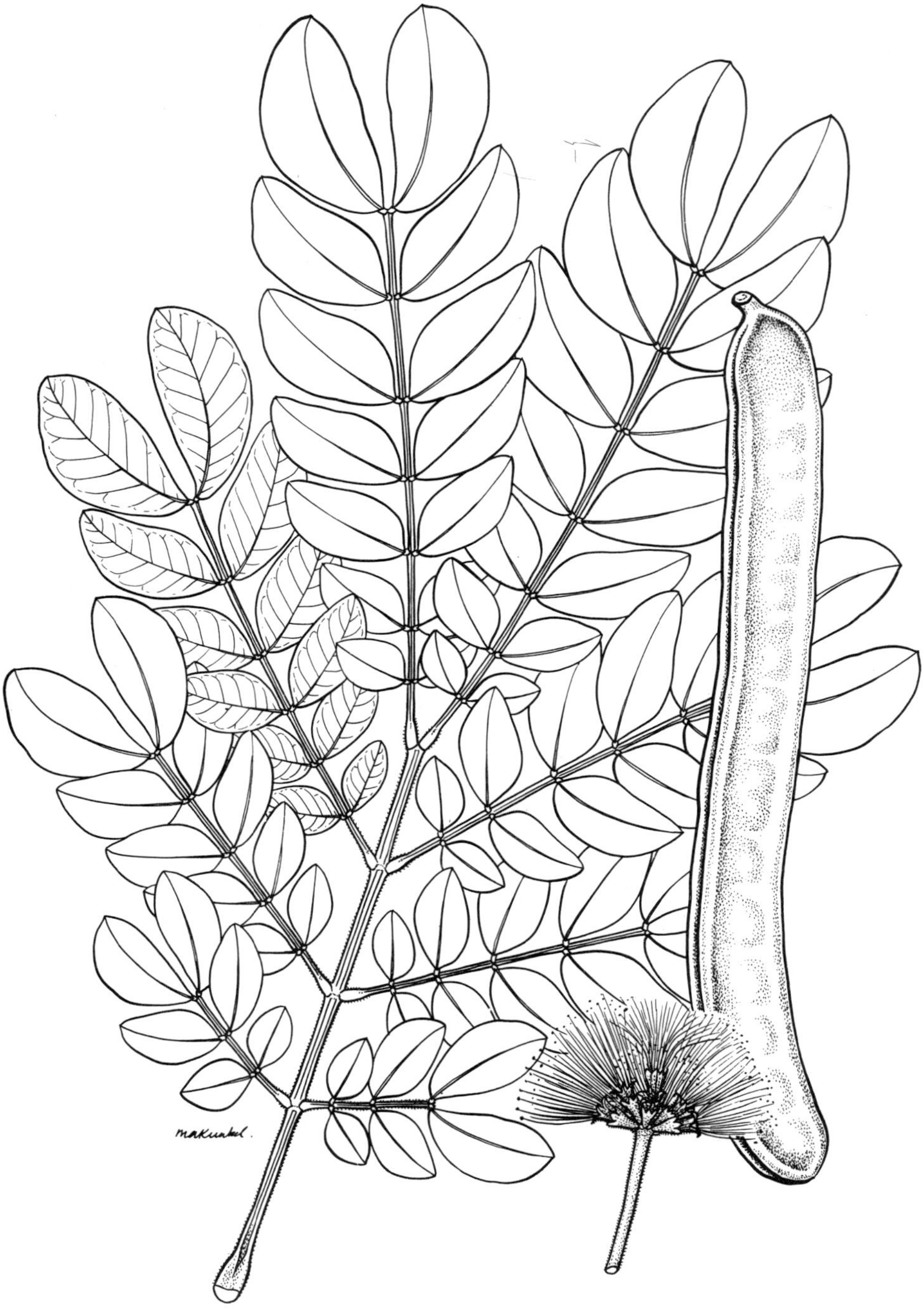

Fig. 77. *Samanea saman*; all drawings = 3/5

CAESALPINIACEAE, the Senna family

The second main group of Leguminosae, with about 150 genera and over 2.000 species, belongs to the Senna or Cassia family. Chiefly trees and shrubs of tropical and subtropical climates. Around 500 to 600 species belong to the genus *Cassia*, another 300 to *Bauhinia*, and about 100 to *Caesalpinia*, the type genus of the family. A number of species are important timber trees of tropical forests, others are cultivated for their ornamental value. In Canary gardens several genera are frequently seen, all of them introduced and well established in this environment.

Bauhinia forficata Link – Orchid Tree

Small tree 6 to 10 m tall, with a high dense or spreading crown. Trunk somewhat bent, up to 30 cm in diameter; bark dark grey, fissured or furrowed; sometimes with basal shoots. New branches prickly, slightly zigzagged. Evergreen (or very briefly deciduous). Leaves alternate, short-stalked, stiff-herbaceous, dark green; blade deeply emarginate (cut more than half way to the base), softly tomentose beneath, up to 15 cm long and 9 cm wide, with two joined glands at the base. Petiole and new shoots also downy. Flower buds usually in pairs, cylindric or finger-shaped. Flowers up to 8 cm across, very showy, with narrow white petals and long, agglomerate, up-curved stamens. Fruit-capsule a woody pod, flat, brownish, up to 10 cm long, usually with remains of the dried flowers at the base; seeds bark brown and shiny.
Propagation from seeds and root-shoots. Best grown in protected gardens at lower altitudes where it makes a fine ornamental. Recommended for roadside planting. The species is native in transitional forest formations of tropical South America. – Several other species are also cultivated in these islands and will be dealt with in the forthcoming volume on 'Flowering Shrubs'.

Ref.: Chittenden; Kunkel 69; Ledin & Menninger; (Lombardo 64?); Menninger 62.

Fig. 78. *Bauhinia forficata*; all drawings = 3/5

Cassia spectabilis DC. – Yellow Shower

Out of the dozen or so mostly shrubby species of *Cassia* grown in these islands, the Yellow Shower – this one lives up to its name – is the most spectacular. This tropical American and West Indian tree is said to reach up to 20 m in height; however specimens in Canary gardens are usually only 5 to 8 meters tall and are pruned annually; new shoots producing growth of up to 3 m during a few months. Trunk short, about 20 cm in diameter; bark dark grey, smooth or only slightly fissured, with distant horizontal grooves. Crown dense, rounded or spreading; young shoots green, slightly tomentose, with lenticels and leaf-scars; brittle-herbaceous and easily breaking off. The species seems to be briefly deciduous. Leaves alternate, paripinnate, 30 to 40 cm long, pendent, on new shoots only, with 10 to 20 pairs of short-petioled, herbaceous, ovate-lanceolate, entire leaflets up to 9 cm long and 2,5 cm wide; blade finely tomentose. Flowers yellow, in very showy, erect, terminal 'spikes' 30 to 50 (80) cm long. Fruit = a cylindrical black pod, up to 25 cm long, about 1 cm in diameter, with numerous horizontal partition walls. Seeds flattened, brown.

Propagated from seeds and woody cuttings, and best grown in cooler places at lower altitudes. It makes a fine garden and roadside tree; unfortunately in these islands its leaves are often infested by a tiny insect commonly known as the 'white fly'. – Botanical synonym: *Pseudocassia spectabilis* (DC.) Britton & Rose.

Several other large species of *Cassia* (e.g. *C.grandis* L.f., the Pink shower, and *C.siamea* Lam., the Kassod Tree) are on trial in these islands but still too small to be considered here.

Ref.: Adams; Barrett 56; Bircher; Corner; Graf; Irvine; Kunkel 69; Little & al.; Menninger 62; (Neal ?).

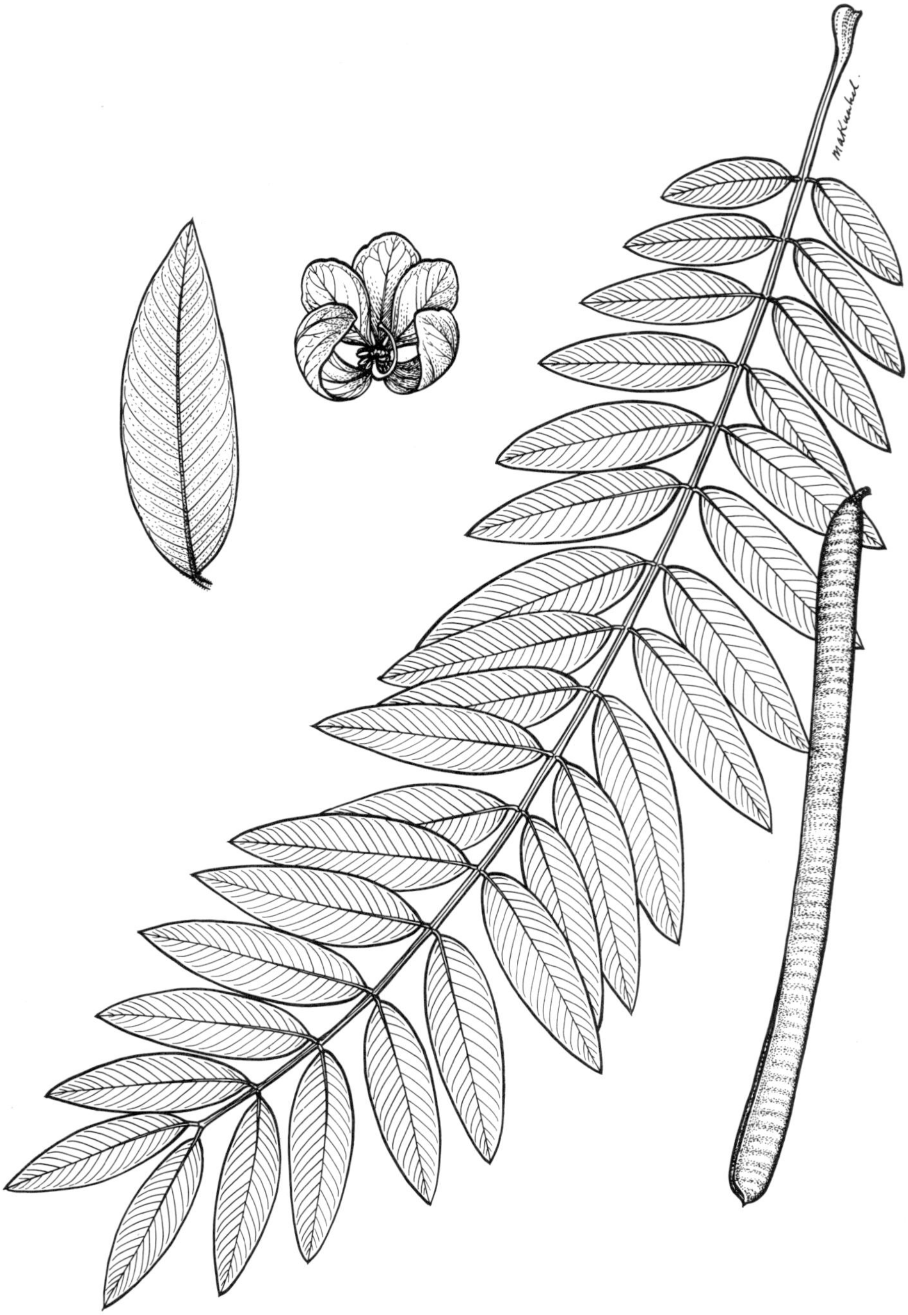

Fig. 79. *Cassia spectabilis*; leaf = 1/2, detail drawings = 3/5

Ceratonia siliqua Linné – Carob Tree

Also known as Locust Tree and St.John's Bread, this species seems to be native in the Eastern Mediterranean region, and is cultivated and/or naturalized elsewhere in subtropical zones. Usually 6 to 10 m high, with a dense, rounded or wide-spreading crown. Trunk short, irregular, 40 to 60 cm in diameter; bark grey, fissured, and flaking off in older specimens. Branches extremely knotty. Evergreen. Leaves alternate, paripinnate, up to 20 cm long, with 3 to 5 pairs of shiny, dark green, coriaceous leaflets; blade ovate, short-petioled, 4 to 6 cm long and 3–3,5 cm wide; apex rounded or slightly emarginate. Flowers very small, yellow-orange to reddish, in cylindrical spikes up to 8 cm long, appearing even on the trunk and older branches (cauliflor!). Fruit = a pod; indehiscent, flattened, hard but elastic, dark brown, 15 to 25 cm long; seeds hard, flattened, shiny brown.

Propagated from seeds and grown in lower, somewhat arid zones; the species is quite resistant to a prolonged drought. Perhaps originally introduced (?) it is now an important tree in Southern Spain where it overtops the native shrub formation. A good shade tree. The fruit pods contain an edible pulp but this is not often used except for feeding cattle and pigs. According to Neal 'in Israel, the average yield per tree is 450 pounds. The flat seeds were the original carat weight of jewelers and apothecaries'. The fruits are rich in carbohydrates and sugar, and a source of alcohol; the ground and roasted seeds serve as a substitute for coffee. The wood is hard and heavy, and is used in carpentry, for wheels and constructions; also for fuel and charcoal (Ruiz de la Torre). Leaves rich in tannin. The industrial uses of a gum obtained from the Carob tree is explained by Esdorn & Pirson.

Ref.: Ball; Barrett 56; Beyron; Bircher; Burkill; Chittenden; Encke; Esdorn & Pirson; Font Quer; Harrison; Kunkel 69; Lombardo 58; Menninger 64; Neal; Perry; Polunin; Polunin & Huxley; Polunin & Smythies; Ruiz de la Torre; Schaeffer; Uphof.

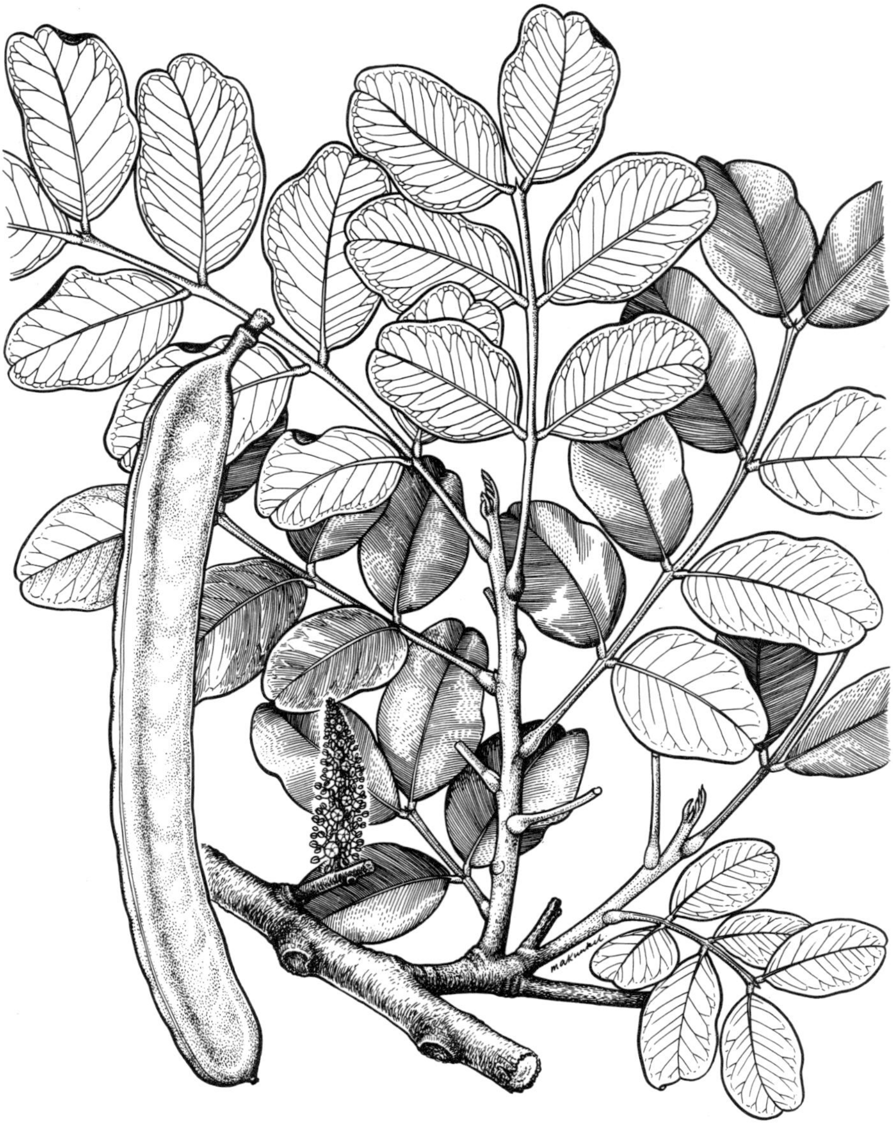

Fig. 80. *Ceratonia siliqua*; drawing = 1/2

Delonix regia (Boj. ex Hook.) Raf. – Flamboyant

Sometimes also called Flame Tree, Peacock Flower, or Royal Poinciana, this ornamental species is native in Madagascar and is cultivated throughout the tropics and subtropics as a shade tree and for its graceful leaves and very beautiful flowers. 'One of the showiest and best-loved trees in Hawaii' (Neal) and surely elsewhere. According to Barrett 'this Flame tree is acclaimed Florida's most spectacular and popular tree', and Corner states that 'this tree is one of the joys of creation'. Flowering whilst leafless, or with the new leaves. At first glance the leaves of the Flamboyant resemble those of a Jacaranda, but they are wider, the pinnae are paripinnate, and the leaflets or segments have a rounded or bluntly pointed apex. The branches are always horizontal and spreading.

Tree 8 to 12 (20) m tall; trunk short, 30 to 50 cm in diameter. Bark grey-brown, smooth, with horizontal grooves and numerous lenticels; splitting or flaking off in older specimens which may develop low buttresses. Foliage deciduous. Leaves alternate, in more terminal position; stalked, bipinnate, oval-shaped in outline, 25 to 40 cm long. Pinnae or primary divisions up to 10 cm long, paripinnate, with up to 20 pairs of oblong, very short-petioled, bright green herbaceous leaflets. Flowers up to 12 cm across; 5 red or crimson petals one being streaked with yellow; very broad-spathulate at the apex and very narrow towards the base. Pods woody, flattened, chestnut-coloured, up to 50 cm long and 4 to 5 cm wide; rather persistant, containing many relatively small dark brown, oblong seeds.

Propagation from seeds; sometimes also successful from woody cuttings. The Flamboyant makes an excellent park and roadside tree. Wood fairly resistant; the bark is said to be a febrifuge, and also contains a gum soluble in water (Burkill). Seeds are made into necklaces. – Common synonym: *Poinciana regia* Boj. ex Hooker. A golden-yellow flowered form has been described as var. *flavida* Stehlé.

Ref.: Adams; Barrett 56; Bircher; Burkill; (Chittenden); Corner; El Hadidi & Boulos; Eliovson; (Encke); Irvine; Keay & al.; Kunkel 69; (Long & Lakela); Menninger 62, 64; Moeller 68, 71; Neal; Perry; Pesman; Schaeffer; Synge.

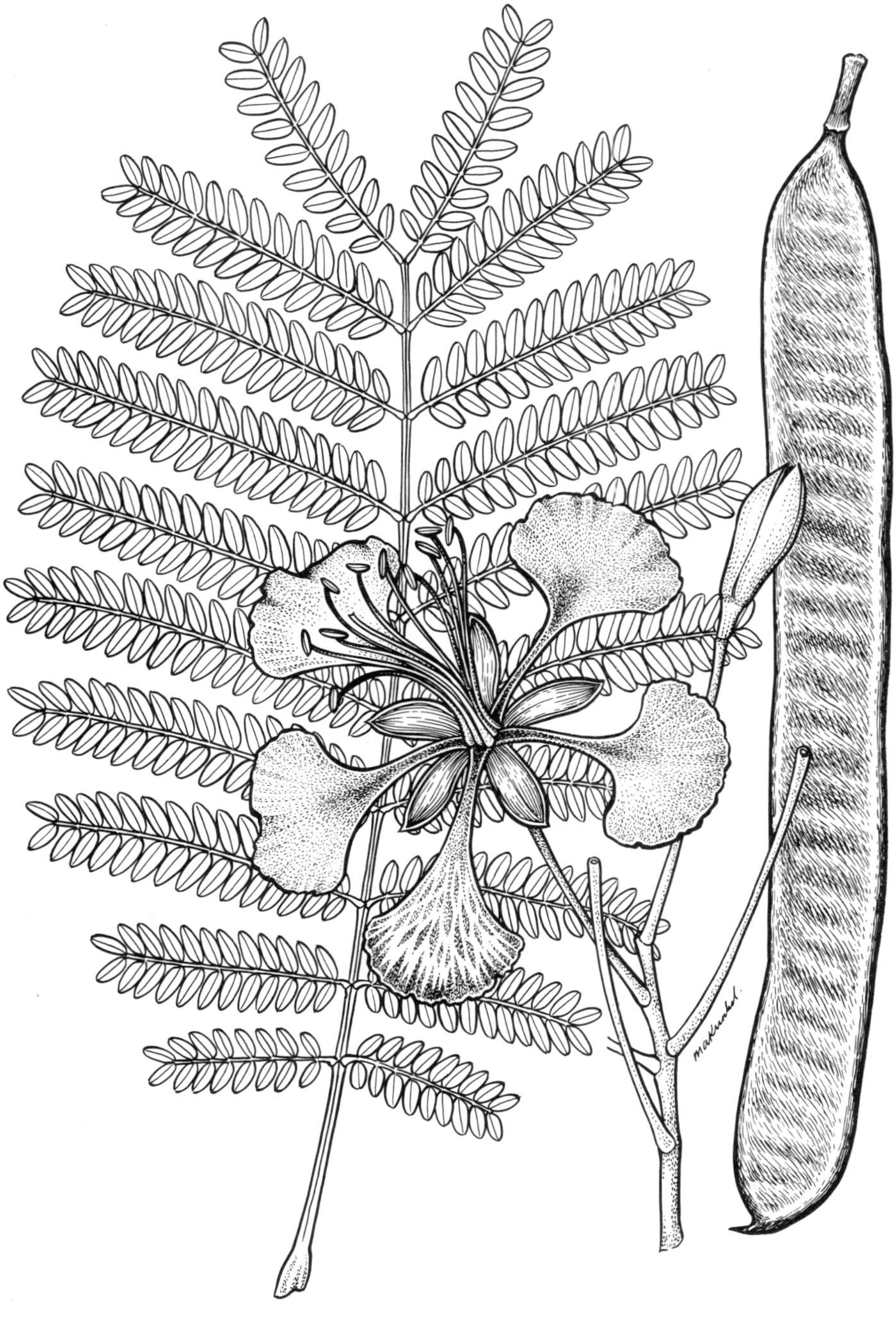

Fig. 81. *Delonix regia*; all drawings = 3/5

Gleditsia triacanthos Linné – Honey Locust

This species is usually seen as a 'weedy' shrub 6 to 10 m tall; however it is said to reach over 30 m in height in mid-southeastern USA, its home country. Frequently with basal shoots which may form new trunks. Crown rounded or spreading, rather open. Bark dark grey to almost black, smooth on younger trees becoming deeply furrowed with scaly ridges on older specimens. Branches heavily armed, often encircled by long, branched spines. Foliage deciduous. Leaves usually paripinnate; young leaves simply pinnate, up to 15 cm long, the following or mature leaves bipinnate, up to 30 cm long. Pinnae 4 to 7 pairs; leaflets herbaceous, oblong, slightly unequal, up to 10 pairs (young leaves with up to 15 pairs). Flowers very small, yellowish-green, in dense racemes up to 8 cm long. Pods flat, reddish-brown, up to 40 cm long and 3,5 cm wide, curved; with hard, greenish-brown, elongated seeds. Propagation from seeds and basal shoots. Usually grown in cooler climates where the species can become an aggressive invader similar to the habit of the *Robinia*; perfectly naturalized in Southern Europe (Ball). Wood very hard and resistant, used for construction, railway sleepers, fence posts, and in general carpentry. Leaves eaten by cattle. – The species is sometimes also cited as *Gleditschia triacanthos* L.; alternative common names: Sweet Bean, Sweet Locust, and Three-thorned Acacia. An almost unarmed cultivar is known as cv. 'Inermis'. The species hybridizes with *Gleditsia aquatica* (= *G.* × *texana*).

Ref.: Ball; Bean; Bircher; Chanes; Chittenden; Encke; Fitschen; Harrison; Lombardo 58; Menninger 64; Mitchell; Polunin & Smythies; Ruiz de la Torre; Uphof.

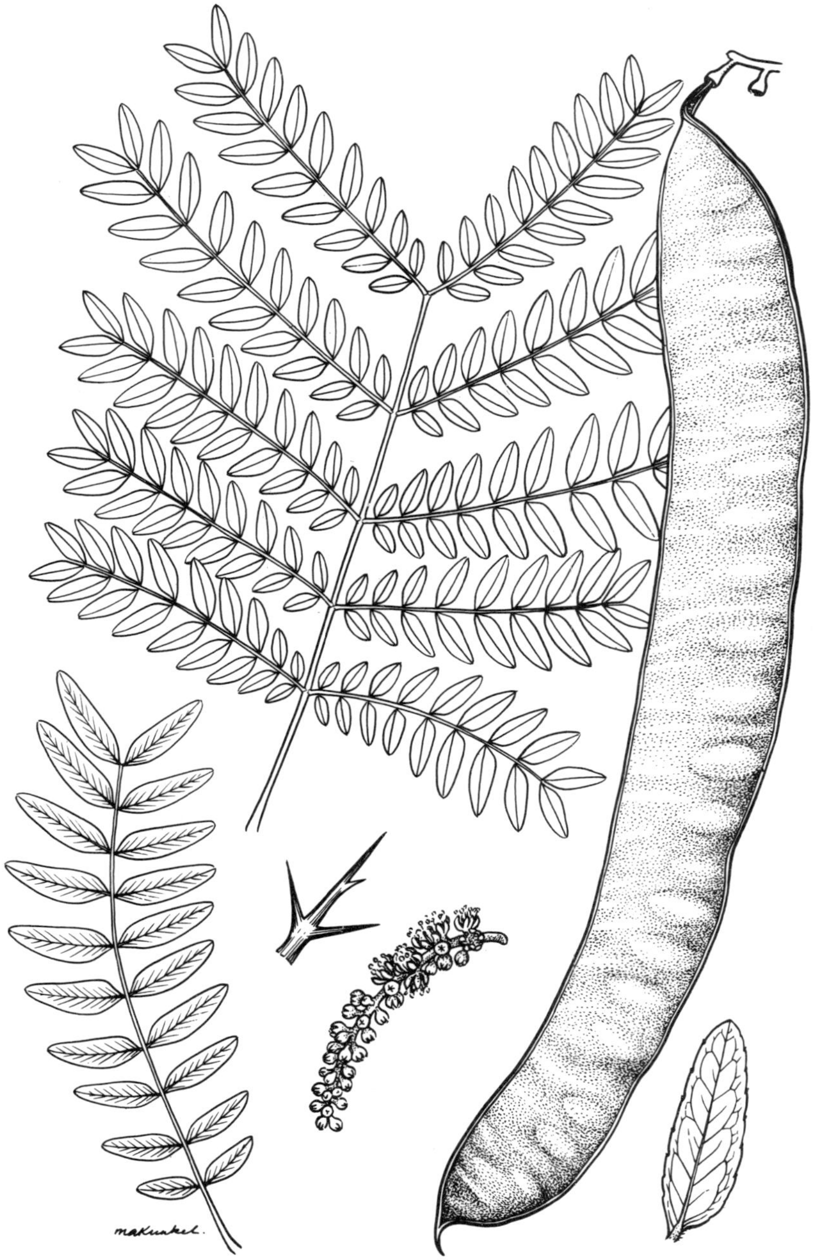

Fig. 82. *Gleditsia triacanthos*; all drawings = 3/5 except solitary leaflet (= about natural size)

Parkinsonia aculeata Linné – Jerusalem Thorn

The common name generally applied to this species is rather misleading because the 'Jerusalem Thorn' is a native of tropical America, and its 'thorns' I would call spines. – Found along drier riversides from Mexico and southern USA to Northern Argentina and Uruguay. Introduced and naturalized elsewhere, including riverbeds of the western Sahara and in the moist savannah of West tropical Africa.

A resistant, spiny but graceful ornamental tree 5 to 8 (10) m tall, with a rounded crown and long pendent branches resembling a 'weeping willow'. Trunk 15 to 25 cm in diameter, bark greenish and smooth; branches conspicuously zigzagged. Foliage deciduous or semipersistant. Leaves bipinnate although this fact is hardly noticeable as each 'leaf' consists of a central rachis converted into a strong spine to which are attached 2 to 3 pairs of pinnae, 30 to 50 cm long; further armament provided by stipular spines. Stalks of pinnae green and able to function as 'leaves'. Leaflets very numerous, oblong or obovate, very small on specimens in arid environments and larger in frequently watered gardens. Flowers in axillary clusters, deep golden-yellow, with a red spotted upper 'lip'. Pods constricted, pale brown, up to 10 cm long; seeds (2–3) oblong, greyish-green and shiny.

Propagation from seeds and woody cuttings; best grown in warmer climates at lower altitudes, fast growing. Foliage and pods browsed by livestock. It is a fine ornamental and also makes an effective hedge. The Jerusalem Thorn is recommended for reafforestation in dry sandy countrysides (Burkill, Irvine). The wood is sometimes used for charcoal. – Alternative common names: Horse-bean, and Paloverde.

Ref.: Adams; Barrett 56; Bircher; Burkill; Chittenden; Eliovson; Irvine; Keay & al.; Kunkel 69; Little & Wadsworth; Lombardo 58, 64; Long & Lakela; Martinez; Menninger 64; Moeller 68; Neal; Pesman; Wiggins & Porter.

Fig. 83. *Parkinsonia aculeata*; flowering branch and details = 3/5

Schotia latifolia Jacq. – Elephant Hedge Bean Tree

A small to medium-sized tree, in Canary gardens only 8 to 10 m tall; trunk 25 to 40 cm in diameter. Bark dark grey, with lenticels and horizontal grooves. Crown rather dense, pyramidal or rounded. Foliage evergreen; however sometimes in the cooler season the appearance of many reddish shiny young leaves is noticed. Leaves alternate, stipulate, paripinnate, up to 10 cm long, with 2 to 3 (or 4) pairs of subcoriaceous, dark green leaflets which are short-stalked, obovate in outline, 4 to 6 cm long and 2 to 3 cm wide. Flowers pinkish, rather showy, in stout terminal panicles up to 12 cm long. Pods flat and woody, brown, almost circular but with a pointed spiny tip. Usually only one large flat, brownish seed; according to Chittenden 1 to 3-seeded.
An admirable ornamental for roadsides and smaller gardens, native in South Africa; propagated from seeds.

A second species, probably *Schotia brachypetala* Sond., quite fast-growing and also native in South Africa, has recently been introduced to these islands but has so far failed to flower. It may turn out to be a good species for planting in more arid zones.

Ref.: Bircher; Chittenden; Kunkel 69; Neal; Palmer & Pitman; Van der Spuy.

Fig. 84. *Schotia latifolia*; all drawings = 2/3

Tamarindus indica Linné – Tamarind

This well-known and widely cultivated species is considered as a native of Southern Asia; its frequent occurrence in tropical Africa may be put down to early introductions by Arab traders. According to Neal the Tamarind is sacred in India and Thailand.

Tree up to 20 (25) m tall, with a rather flat, wide-spreading crown. Trunk somewhat bent, or twisted; bark greyish-brown, fissured, becoming ridged and corky in older specimens. Foliage evergreen; young twigs drooping. Leaves alternate, paripinnate, usually in double rows, about 10 to 15 cm long, with 12 to 20 pairs of somewhat sensitive leaflets; they tend to fold up in overcast weather often a sign of rain (Menninger 67). Leaflets short-stalked, oblong but unequal, herbaceous, dark green, up to 2 cm long. Flowers showy, with cream-coloured sepals and red-veined petals; in terminal panicles with few flowers each. Pods flattened, somewhat constricted, brown, slightly velvety, up to 15 cm long; seeds flat, rather large.

Propagated from seeds and woody cuttings, this species makes an excellent shade tree at lower altitudes. The pulp of the fruit is edible and used in jellies, deserts, sauces, pickles, beverages and refreshing drinks. The juice is slightly laxative and used in medicine (Barrett, Burkill, Irvine). According to Uphof it is antiscorbutic. The wood is used in carpentry, for fuel, and makes a charcoal used in gunpowder. Irvine reports that fresh flowers are eaten in Nigeria; the leaves are browsed by cattle and are also eaten as vegetables in India (Burkill). The bark contains tannin and a dye; twigs are used as chewsticks.

Ref.: Adams; Barrett 56; Bircher; Burkill; Calabria; Chittenden; Corner; Encke; Graf; Irvine; Keay & al.; Kennard & Winters; Kunkel 69; Little & Wadsworth; Long & Lakela; Martinez; Menninger 64, 67; Molesworth Allen; Neal; Purseglove; Uphof.

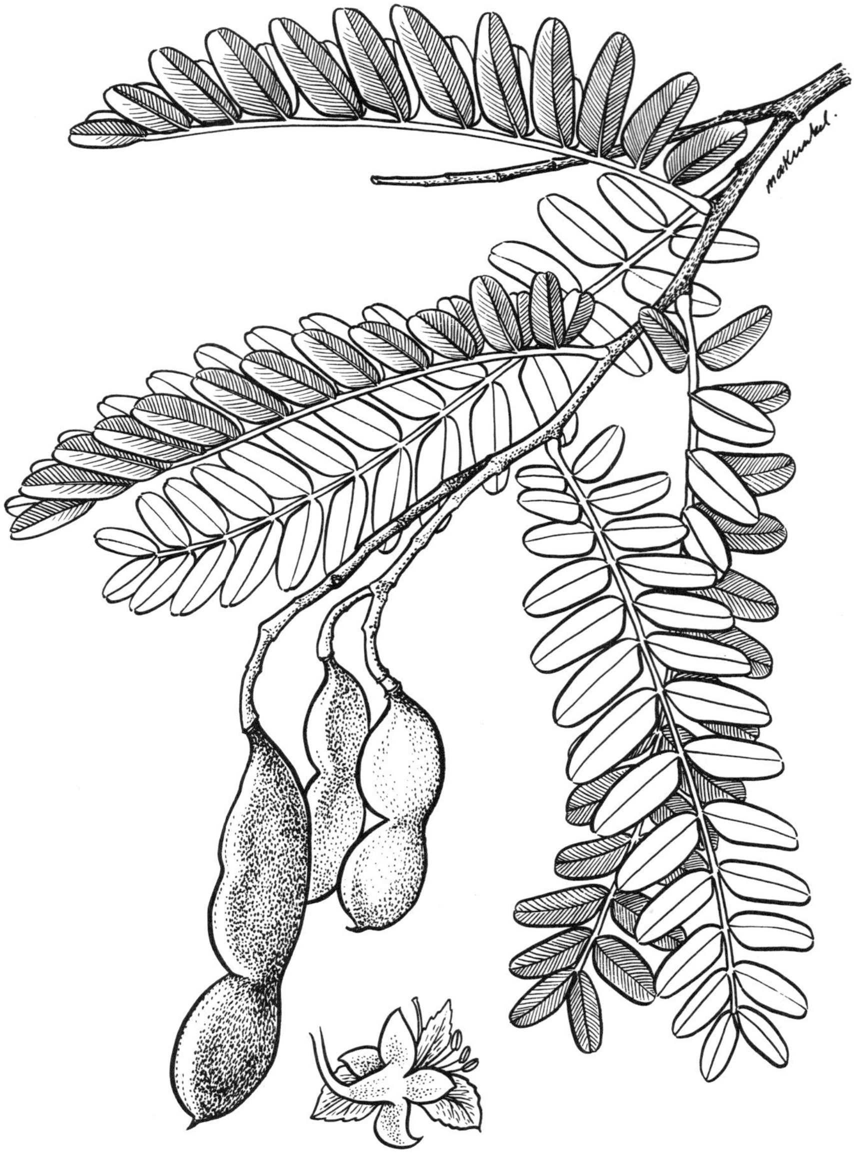

Fig. 85. *Tamarindus indica*; drawings = 3/4

FABACEAE, the Pea family

This last and most varied family of the Leguminosae is distributed all over our living planet. As a rough guess: about 400 genera with some 8.000 species of which 2.000 already belong to the genus *Astragalus*. Plants of economic importance include species of *Arachis* (Pea-nut), *Cicer* (Chick-pea), *Glycine* (Soja-bean), *Indigofera* (Indigo), *Lens* (Lentil), *Phaseolus* (Bean), *Pisum* (Pea), *Vicia* (Alfalfa) etc.; *Laburnum*, *Dalbergia*, *Genista*, *Hardenbergia*, *Sophora* and others are widely cultivated ornamentals, and *Ulex* has become a noxious weed in some places. – There are about 170 species of Fabaceae in woods and fields of the Canary Islands, some 40 of them are considered as endemics.

Erythrina caffra Thunb. – Lucky-bean Tree

A South African species also known as Coral Tree, and Kaffierboom, highly decorative, drought-resistant, and much appreciated for larger parks and gardens. Usually 8 to 12 m tall, according to Palmer & Pitman up to 20 meters in its native forests. Trunk short, 40 to 60 cm in diameter; bark grey, fissured. Wide spreading flattened crown with very thorny branches; branches easily breaking off. Foliage deciduous. Leaves alternate, imparipinnate, long-stalked, leaving scars on younger branches; with 3 (to 5) leaflets of which the terminal one is the largest, up to 8 cm wide; blade triangular-rhomboid, glabrous, herbaceous, pointed at apex, and glandular at the base. Flowers red to scarlet, with waxy petals in very showy, usually erect, terminal clusters or elongated racemes; flowering whilst leafless. Fruits (pods) constricted and curved, few-seeded, up to 12 cm long. Seeds reddish and shiny.
Propagation from seeds and cuttings, and best grown at lower altitudes. The wood is light and spongy, used for canoes and to make floats for fishing nets. The seeds are poisonous but made into necklaces and are worn as 'lucky beans'.

Ref.: Chittenden; (El Hadidi & Boulos); (Eliovson ?); Graf ?; Harrison; Kunkel 69; Menninger 62; McClintock; Palmer & Pitman; Perry; Synge; Van der Spuy.

Fig. 86. *Erythrina caffra*; all drawings = 3/5

Erythrina crista-galli Linné – Cock's Comb Coral

Another fairly common Coral tree, of South American origin and now widely cultivated in all suitable climates. According to Lombardo up to 20 m tall; in gardens of these islands usually 4 to 6 m in height, with a short and bent trunk 20 to 30 cm in diameter. Bark yellowish-brown, thick, furrowed becoming corky. Crown spreading, open, irregular. Foliage deciduous. Leaves and flowers on each year's new branches. Leaves trifoliate, herbaceous, long-stalked, armed; leaflets shiny green, ovate-elliptic, up to 8 cm long and 5 cm wide. Flowers very deep red or crimson-scarlet, 3 to 5 cm long, in very showy terminal inflorescences up to 50 cm long. Pods constricted, brownish, 10 to 15 cm long, with large shiny brown seeds.

Propagation from seeds and cuttings. Best grown in warm humid climates; in drier zones the terminal branches usually dry off. A highly esteemed species suitable for the smaller garden. Wood very soft. Bark used medicinally (Little & al.). – Alternative name: Lobster plant (Bircher).

Ref.: Beyron; Bircher; Chanes; Chittenden; Corner; Eliovson; Encke; Graf; Harrison; Kunkel 69; Little & al.; Lombardo 58, 64; McClintock; Menninger 62; Moeller 68, 71; Neal; Perry; Synge.

Fig. 87. *Erythrina crista-galli*; flowering branch and fruit = 2/5

Erythrina variegata Linné – Indian Coral-tree

Often cited as *E.indica* Lam., and also known as Immortelle, Tiger's Claw and Indian Coral-bean, this species is native in tropical Asia and some Pacific archipelagos. A small tree 5 to 8 m tall, inconspicuously armed, with rather large leaflets (10 to 15 cm long) and a relatively long and narrow inflorescence (20–30 cm) with many narrow scarlet flowers. The terminal leaflet is often bent at the base. The scarce material found in Canary gardens probably belongs to var. *orientalis* (L.) Merr. which is grown in some countries as a shade tree for crops; according to Menninger the leaves are used as a green manure. Uphof reports that in Siam (Thailand) the white wood of this species is made into a powder and used as face-powder. Leaves, bark and flowers have medicinal properties. According to Huxley (p. 138) these 'Coral Trees, which make so much nectar that they are called 'Cry-baby Flowers', are widely visited by birds in the dry season when they bloom'. Perry says that 'the young leaves are eaten in curry and in Indo-China used for wrapping minced meat'.

Ref.: Adams; Barrett 56; (Bircher); Burkill ?; (Chittenden); (Corner); (Graf); (Harrison); Huxley; Little & al.; McClintock; Menninger 62; Neal; Perry; Synge; (Uphof).

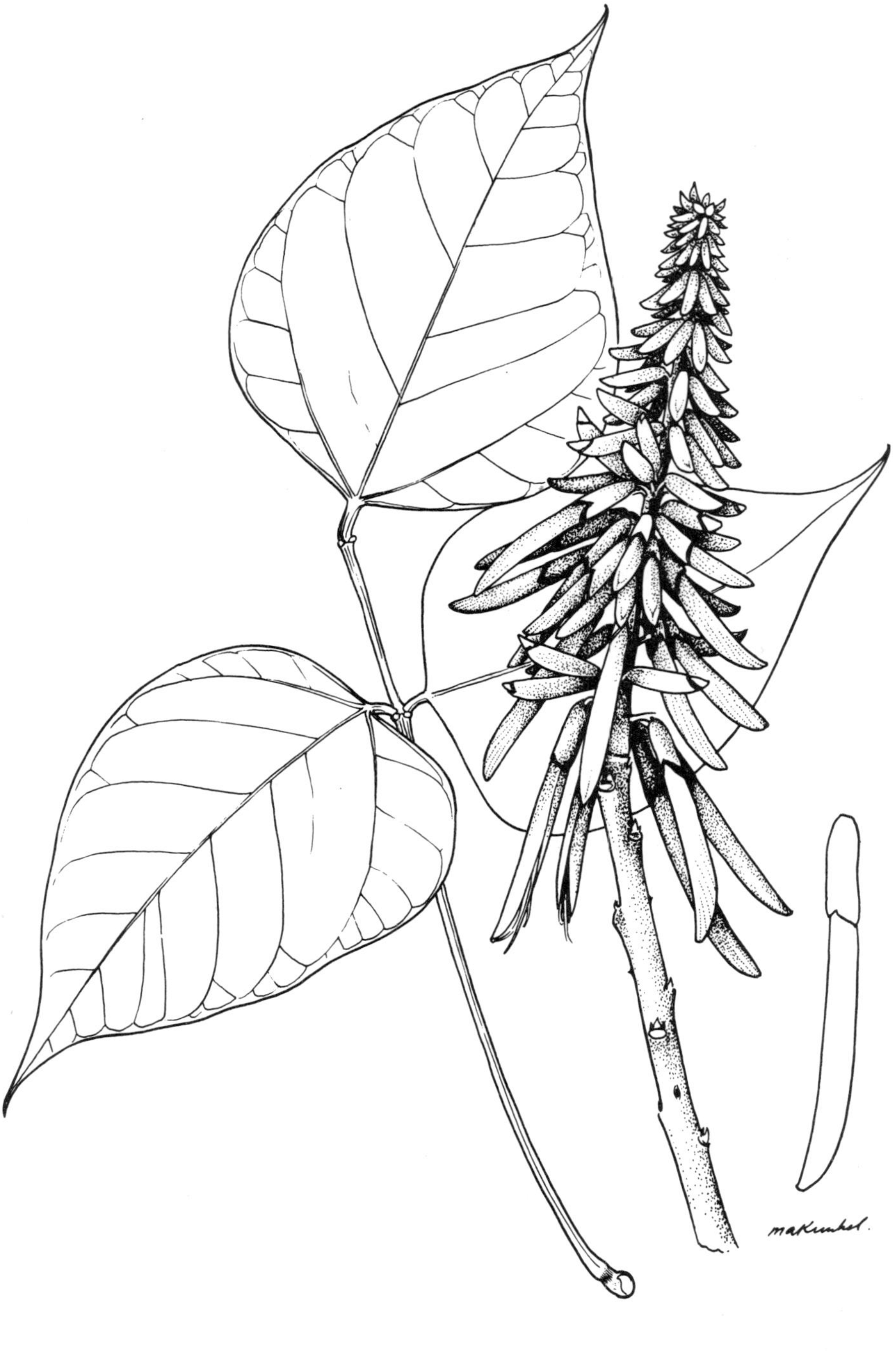

Fig. 88. *Erythrina variegata*; all drawings = 1/2

Robinia pseudacacia Linné – False Acacia

Used here is the spelling as given by P. W. Ball in 'Flora Europaea', although Linnaeus (1753) described the species as *Robinia pseudo Acacia* which, according to the 'Willis Dictionary' reads *R.pseud-acacia* L. Other authors (i.e. Chanes, Chittenden, Fitschen) refer to the same species as *R.pseudoacacia* L.; another version (viz Neal) reads *R.pseudo-acacia* L. And Linnaeus died 200 years ago, unable to defend the original name! – But we have to agree to one name, even if it seems incorrect, also bearing in mind that the plant we refer to is also known as Black Locust, Common Locust, and Acacia. Perry says it is a 'noble tree from North America'. I try not to be discriminate but I remember the species as being rather a nuisance on sandy ground in Central Europe. However it is a tree up to 25 m tall which permits cutting, grafting and other treatment, having a slender trunk, a deeply furrowed bark and a spreading crown. Branches and twigs with spiny stipules; the tree is leafless in the European winter. It produces very fragrant flowers attracting bees, and is able to grow in poor ground already given up by other woody vegetals. Leaves alternate, imparipinnate, 20 to 30 cm long; leaflets up to 21, herbaceous, oval-shaped, dark green above and paler beneath. Flowers white, in occasionally upright but usually pendulous racemes 10 to 20 cm long. Pods flat, brown, 6 to 10 cm long; seeds black and shiny.

Propagation from seeds and woody cuttings. It makes a useful hedge, when regularly pruned also a good street tree. The False Acacia is a valuable species for reclamation of eroded slopes. Garden specimens usually chosen from an unarmed variety. Wood hard and durable, used for constructions, ship-building, and for furniture. According to Uphof the species has poisonous properties but cooked seeds are edible. Fresh flowers are also edible, and are the source of a delicious honey. Leaves are eaten by livestock.

Ref.: Ball; Chanes; Chittenden; Eliovson; Encke; Fitschen; Harrison; Kunkel 69; Lombardo 58; Neal; Polunin & Huxley; Ruiz de la Torre; Uphof; Willis.

Fig. 89. *Robinia pseudacacia*; drawing = 3/5

Tipuana tipu (Benth.) O.Kuntze – Tipuana

Cited by several authors as *T.tipu* (Benth.) Hubbard & Rehder, this admirable roadside tree is little known and often confused with so-called related species.

Of South American origin, this species grows up to 20 m in height, has a short but pronounced trunk with a dark grey, furrowed bark, and a dense and wide spreading crown. Briefly deciduous (evergreen according to Menninger), with large imparipinnate leaves up to 25 cm long. Leaflets oblong, herbaceous, dark green, 3 to 5 cm long. Flowers yellowish-orange, in showy terminal bunches. The fruit, which is winged and looks like one half of a Maple fruit (samara), reminds us that this monotypic, somewhat neglected member of 'legumes' is in need of further investigation.

Sometimes also given as *Tipuana speciosa* Benth., and as Tipa or the Pride of Bolivia. It is a frequent avenue-tree in Funchal (Madeira), and has only recently been introduced to the Canary Islands. Propagation from seeds; a fast growing species which thrives best at lower altitudes.

Ref.: Eliovson; Graf; Lombardo 58?; Menninger 62; Neal.

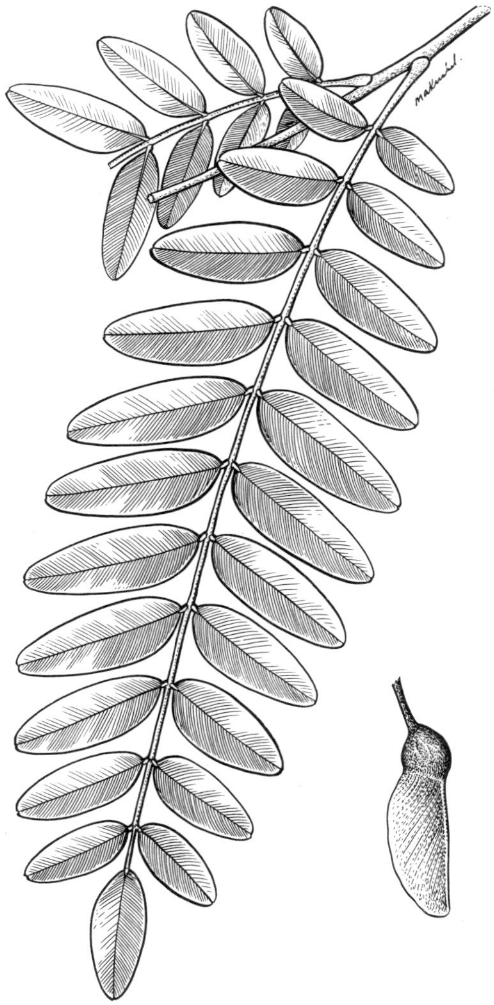

Fig. 90. *Tipuana tipu*; leaf and fruit = 3/5

LYTHRACEAE, the Crape Myrtle family

A widely distributed family of herbs, shrubs and trees consisting of about 25 genera with over 500 species. Some weedy species (*Lythrum*) are of cosmopolitan distribution. The genus *Cuphea* (mostly subshrubs) is widely cultivated.

Lagerstroemia speciosa (L.) Pers. – Crape Myrtle

A medium-sized ornamental tree native in Southeast Asia and (?) Australia where it may attain a height of about 30 meters. In cultivation usually 8 to 12 m high, with a trunk 25 to 40 cm in diameter. Bark greyish-brown, somewhat fissured or scaly. The foliage is deciduous making a colourful autumn display. Leaves mainly opposite so arranged that they appear to have been flattened; petiole short, blade oblong, subcoriaceous, green but tinged with purple, 12 to 15 (25) cm long and up to 5 cm wide. Flowers pink or mauve, in large terminal panicles; curiously shaped, with very showy, wrinkled spathulate petals. Each flower up to 7 cm across. Seed capsule woody, roughly spherical, up to 2,5 cm in diameter containing numerous flat, winged seeds.
This tree is cultivated as an ornamental. However in its home country an important timber tree; the wood is used for railway sleepers, in shipbuilding, paneling and carpentry. It is frequently used as a substitute for walnut wood, and is said to be resistant to attacks by insects. Hegi reports that the wood is as much appreciated as Teak. Irvine calls it 'one of the most beautiful of tropical trees' and says that it is naturalized in West Africa. Most parts of the tree are used in local medicine; the bark contains tannin, and an infusion of the bark may cure diarrhoea (Burkill). Menninger and Adams call the species the 'Queen's Flower Tree'; it is the 'Queen Crape myrtle' of Bircher, and the 'Rose of India' of Corner. Eliovson calls it the 'Pride of India', whereas Harrison spells the name 'Crepe Myrtle', and cites a hybrid with *L.indica* L., a smaller leaved, shrubby but otherwise much related species also grown in some Canary gardens. – A frequently used synonym is *Lagerstroemia flos-reginae*.

Ref.: Adams; Barrett 56; Bircher; Burkill; Chittenden; (Corner); Eliovson; Harrison; Hegi V/2; Irvine; Little & Wadsworth; Menninger 62; Neal; (Uphof).

Fig. 91. *Lagerstroemia speciosa*; flowering branch and fruit = 3/5

COMBRETACEAE, the Combretum family

A family of trees, shrubs and climbers, with about 20 genera and some 600 species; mainly in the tropics. Species of *Combretum* and *Quisqualis* (the Rangoon creeper) are ornamental vines, and the genus *Terminalia* provides large valuable timber trees. *Laguncularia* is a member of the Mangrove forest. More common in cultivation only

Terminalia catappa Linné – Indian Almond

Also known as the Tropical Almond, this species is usually a small or medium-sized tree best cultivated in parks and gardens near to the sea. It is probably native in Southeastern Asia and Oceania where the species is described as a large tree; Fosberg et al. give it a pantropical distribution but this may refer to its present occurrence. Wiggins & Porter, the authors of the 'Flora of the Galápagos Islands' believe the species to be native in the Indo-Malaysian and Polynesian region. It is listed as introduced and widely naturalized in many American and African 'Floras'.

Usually up to 15 m tall, with a short trunk 30 to 60 cm in diameter; bark dark grey and somewhat fissured. Branches slender, spreading horizontally in a kind of verticillate (whorled) arrangement. Crown large, flattened or pyramidal, according to the particular conditions of its habitat. The species is known as an evergreen in humid climates but is definately deciduous in more arid zones; however, according to Burkill this also applies to Southeast Asia because 'its leaves turn red before they fall, and then make the tree very conspicuous'. Leaves more or less in whorls in terminal position, short-stalked, leathery or nearly so, glossy, obovate to almost spathulate measuring 15 to 20 by 6 to 10 cm. Leaves of young plants much larger, purplish and shiny. Flowers small, whitish, in terminal spikes. Fruits somewhat almond-like, with narrowly winged edges, light brown when ripe, 5 to 7 cm long.

The wood is used for construction and boat-building; bark, roots and leaves have medicinal applications, and the seeds are eaten raw or roasted. The bark and fruits provide a black dye which, according to Uphof, is used in parts of India for colouring teeth. The oil obtained from the seeds is apparently of good quality similar to the oil of real almonds. The Indian Almond makes a good street tree, and is recommended for sea-shore plantations. It is resistant to salt spray as well as to occasional droughts if not too prolonged.

Ref.: Adams; Barrett 56; Bircher; Burkill; Chittenden; Degener; Encke; Fosberg et al.; Irvine; Keay et al.; Kunkel 69; Little & Wadsworth; Long & Lakela; Menninger 64; Pesman; Uphof; Wiggins & Porter.

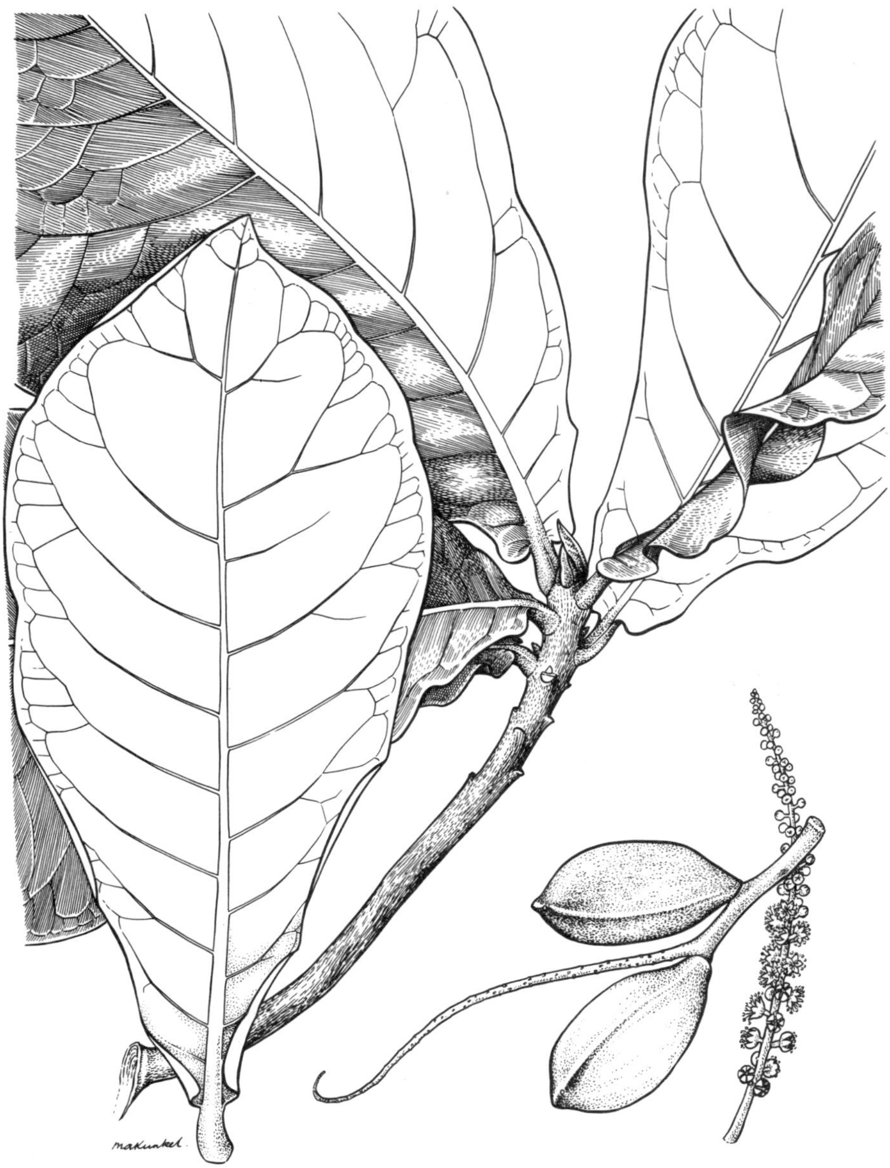

Fig. 92. *Terminalia catappa*; all drawings = 1/2

MYRTACEAE, the Myrtle family

This large family mainly of trees and shrubs shows two principal distribution centres: Australia and America, and is found in both tropical and subtropical climates; relatively few genera enter more temperate zones. Chittenden and Neal, for example, say that about 70 genera and 2.800 species belong to this family, whereas Willis gives a round number of 100 genera producing berries or capsules, drupes or nuts. Many species are cultivated either for their fruits, as timber trees, or for their ornamental value. *Myrtus*, *Callistemon*, *Eugenia* and *Feijoa* are large ornamental shrubs or small trees.

Eucalyptus spp. – The Eucalypts

After the now generally accepted separation of *Eugenia* aggr. into several genera, *Eucalyptus* remains the largest genus within this family. Willis gives 500 species for the genus, Neal keeps the number limited to about 300, whereas Ehrendorfer and Menninger mention about 700 species. The new 'Classification of the Eucalypts' by Pryor & Johnson limits the genus to about 400 species and 60 subspecies. According to R. C. Barnard (in Synge) 'over 100 of the described species are known to be hybrids'. The genus is chiefly native in Australia and Tasmania, with a few species also in New Guinea and neighbouring islands. Usually planted for reafforestation and as a source of fuel and wood pulp. No species of *Eucalyptus* is recommended for countrysides characterized by chronic watershortage. – There are about two dozen different Eucalypts grown in the Canary Islands; however most are rare, and only the following have been selected.

Fig. 93. *Eucalyptus camaldulensis*; all drawings (including juvenile leaves, right) = 3/5

Eucalyptus camaldulensis Dehnhardt – Red Gum

Tree up to 20 m tall but considerably larger in its home country. Trunk high, usually somewhat twisted; bark grey, continually scaling off in larger sheets. Crown high and spreading, outer branches drooping. Evergreen, as are all the Eucalypts mentioned here. Leaves alternate, stalked, hanging, coriaceous, greyish-green; juvenile leaves oblong-lanceolate, adult leaves linear-lanceolate, slightly curved, up to 20 cm long. Buds in stalked clusters, usually 6 to 10 on a common stalk, small, with a pointed operculum. Calyx-tube bowl-shaped. Numerous short, whitish or greenish-white stamens. Fruit-capsule small, disk-like above and bowl-shaped beneath; disk 5 to 8 mm across. Seeds very small; according to Goor & Barney there are between 200.000 and 1.000.000 seeds per kg, depending on the provenance of the material.
Native in Australia in general. Synonym: *E.rostrata* Schlecht.; alternative common names: Murray Red Gum, and Australian Kino. – According to Uphof its red gum is used medicinally for coughs and colds, dysentery, and relaxed throats; it contains kinotannic acid and a glucoside. The very hard and durable wood is used for fence posts, shipbuilding, bridges, railway sleepers etc., also for cellulose and wood pulp. Fast growing tree propagated from seeds; usually seen along roadsides. In these islands in almost every climatic zone but truly naturalized in cooler, more cloudy ravines. The Red Gum is often recommended for afforestation in arid zones (see Goor & Barney, Kaul & al., etc.).

Ref.: Barrett 56; Bircher; Burges; (El Hadidi & Boulos); Goor & Barney; Kaul & al.; Kunkel 69; Lombardo 58; Moggi; Neal; Polunin & Smythies; Ruiz de la Torre; Uphof.

Eucalyptus cornuta Labill. – Horned Gum

Medium-sized tree with a fairly straight trunk and a persistant, furrowed and very dark bark. Crown spreading. Leaves stalked, lanceolate, dark green, leathery, up to 12 (14) cm long and 2 (2,5) cm wide; young leaves opposite, almost orbiculate. Adult specimens are easily recognized by their horned buds and fruit-capsules. Flower buds (with a lid, or operculum) 8 to 15, clustered on a slender stalk 2,5 to 3,5 cm long, buds themselves very short-stalked, long-horned (4–5 cm long), and somewhat curved. Flowers (stamens) very showy, greenish-white, in dense clusters up to 8 cm across. Fruit-capsules more or less bell-shaped and strongly horned, clustered on a long slender stalk; spiny horns up to 1,5 cm long.
A roadside tree not too frequent in these islands. Propagation from seeds. Growing for preference in zones known as the 'cloud belt'. The species is of Western Australian origin. Its wood is said to be very hard but elastic, and serves for general building purposes.

Ref.: Bircher; Chittenden; Lombardo 58.

Fig. 94. *Eucalyptus cornuta*; drawings = 2/3

Eucalyptus ficifolia F.v.Mueller – Scarlet-flowering Gum

This rather spectacular species (whilst flowering) is found in some parks and gardens, and is one of the few really appreciated members of the genus, at least as far as its ornamental value and non-destructive characters are concerned. The Scarlet-flowering Gum is native in SW Australia.
Small tree up to 15 m in height; trunk short and slender; bark persistant, dark grey and furrowed. Leaves petiolate, broad-lanceolate, rather thick, dark green, about 12 cm long and 4 to 5 cm wide, with fine lateral veins; midrib yellowish. Bright scarlet or crimson flowers (somewhat waxy stamens) in showy terminal corymbs. Buds short, stalked, with a short-pointed operculum. Fruits woody, cup- or urn-shaped, about 2 cm in diameter.

The species is best grown at lower altitudes and seems susceptible to low temperatures. Propagation from seeds. According to Harrison in its home country much attacked by a serious disease which affects the bark causing the eventual collapse of the tree.

Ref.: Bircher; Chittenden; Eliovson; Graf; Harrison; Lombardo 58; Menninger 64; Moeller 68, 71; Perry.

Fig. 95. *Eucalyptus ficifolia*; drawings = 2/3

Eucalyptus globulus Labill. – Blue Gum

One of the largest and perhaps the best known species of the genus, grown as a roadside tree or in large plantations. Native in South Australia and Tasmania. This species and other members of the genus seem to fail in very humid tropical countries; otherwise it grows from sea-level to over 3.500 meters (S. American Andes). Because of its enormous root-system the Blue Gum can be rather destructive when growing near to buildings, it lifts up paving stones and breaks waterpipes and channels. Freshly cut stumps usually produce exuberant new basal shoots.

In these islands about 25 to 30 meters tall; however the Blue Gum is said to attain a height of 100 meters in Australia. Trunk corpulent (1–1.5 m ∅), somewhat twisted; bark short-lived coming off in large flakes or ribbons. Crown very large, almost pyramidal. Foliage bluish-green and extremely heterophyllous: juvenile leaves slightly sticky, opposite, sessile, cordate-ovate, herbaceous, 4 to 15 cm long and 4 to 8 cm wide; adult leaves alternate, lanceolate and somewhat falcate, stalked, coriaceous but brittle, up to 25 (30) cm long and 2 to 4 cm wide. Both forms strong smelling. Flowers large, solitary or up to three together, axillary, almost sessile; calyx bell-shaped, operculum warty and covered by a bluish-white wax; open flowers (white stamens) up to 5 cm across. Fruit-capsule roughly bell-shaped but angled, 2 to 3 cm in diameter.

The Blue Gum is quite hardy and also drought-resistant; it is much recommended by specialists for afforestation in arid zones (e.g. Kaul & al.). Timber very strong, used for shipbuilding, handles, wheels, fuel and wood pulp. Posts and poles often used in the mining industry; according to Flinta (in Kaul & al.) 'miners in Bolivia and southern Peru demand the use of *E.globulus* in the galleries because this timber warns before eventual failures'. – The leaves of the Blue Gum are the source of Oleum Eucalypti, an oil having many properties: antiseptic, expectorant, febrifuge, antiperiodic etc. (Uphof, Font Quer, Martinez). Young specimens are often grown as potplants. The species propagates easily from seeds but freshly cut branchlets used as fence posts may also strike and thrive. 'Eucalyptus globulus compactus', the Bushy Blue Gum of Graf, according to Pryor & Johnson seems to be a hybrid between *E.globulus* and a still undetermined species. Alternative names of this tree: Fever Tree and Tasmanian Blue Gum.

Ref.: Bean; Beyron; Bircher; Burges; Burkill; Chanes; Chittenden; Encke; Font Quer; Graf; Kaul & al.; Kunkel 69; Lombardo 58; Martinez; Mitchell; Moeller 68; Moggi; Neal; Perry; Pesman; Polunin & Smythies; Ruiz de la Torre; Schaeffer; Soukup; Uphof.

Fig. 96. *Eucalyptus globulus*; all drawings (including juvenile leaf, left) = 3/5

Eucalyptus lehmannii (Preiss. ex Schau.) Benth. – Bush Yate

The common name 'Bush Yate' is also sometimes given to *E.cornuta*, a closely related species but long-leaved and forming a real tree whereas *E.lehmannii* is usually a very irregular-shaped large shrub or small tree, with short oval-shaped or obovate leaves. The Bush Yate is grown along roads and in gardens, and is quite a sight when in full bloom. The species is native in Western Australia.

Specimens usually 5 to 7 m tall, with a short, much twisted trunk; bark reddish-brown and very scaly. Crown rounded and spreading. Branches fissured or scaly, with woody persistant fruit-bunches from previous year's growth. Leaves short-stalked, coriaceous, dark green, up to 7 cm long and 2 cm wide, with a mucronate apex; blade with oily glands showing against the light. Buds, flowers and fruits as in *E.cornuta* except all are sessile on a stouter and shorter common stalk. – Propagation from seeds; rather slow-growing.

Ref.: Harrison; Kunkel 69; Menninger 64; Moeller 68.

Fig. 97. *Eucalyptus lehmannii*; all drawings = 3/4

Eucalyptus occidentalis Endlicher – Swamp Yate

Medium-sized tree from Western Australia making an effective roadside tree in these islands. Trunk fairly straight; bark persistant but rather stringy, dark grey. Crown large and dense. Juvenile leaves usually already alternate, ovate-lanceolate; adult leaves alternate, spaced, lanceolate and often falcate, coriaceous, up to 12 cm long and 3 cm wide, narrow pointed. Common flower- (and fruit-) stalks up to 3 cm long; 5 to 8-flowered. Buds narrow, campanulate with a very short-horned operculum up to 2 cm long; individual buds (and fruits) stalked. Yellow-flowered. Fruit-capsule bell-shaped, woody, up to 1 cm in diameter.

Propagation from seeds. According to Goor & Barney 'the wood contains 15–25 per cent tannin and is used in construction'. The Swamp Yate is frequently planted in saline environments, especially in the Middle East (Kaul & al.).

Ref.: Goor & Barney; Kaul & al.; Lombardo 58; Moggi.

Fig. 98. *Eucalyptus occidentalis*; fruiting branch and details = 3/5

Metrosideros excelsa Sol. ex Gaertn. – New Zealand Christmas Tree

This large and very impressive tree, the Pohutukawa of the Maoris (a much nicer name!), is called by Menninger (1964) the 'largest Christmas tree in the world'. There are larger ones, however, its common name is connected with its flowering season as around Christmas the tree is covered with masses of bright crimson flowers. – Frequently used synonymous name: *M.tomentosa* A. Rich.

Tree up to 20 m tall, with a dark and dense, spreading crown. Trunk short and heavy, often bent and soon ramified; bark thick, dark grey, rough, deeply furrowed and coming off in large strips. Evergreen. Leaves opposite, decussate, arranged more or less in two rows; petiolate, oblong, coriaceous, the upper surface being dark green, the under side pale green and pubescent; 5 to 8 cm long and 2 to 3,5 cm wide. Nerve system very fine, with a noticeable false marginal vein. The numerous crimson stamens about 2 to 3 cm long; several flower-heads together in terminal position forming showy false umbels. Fruit-capsules obconic or bell-shaped, tomentose, 3-valved, with numerous very small chestnut-coloured seeds.

Propagation from seeds, cuttings and layers. The species thrives best in cooler zones of lower altitudes. In its home country said to be a coastal tree. According to Menninger (1962) the tree may produce aerial roots. Uphof (curiously enough he cites both *M.excelsa* and *M.tomentosa* as different species) states that the timber is compact and very durable. An infusion of the bark is used for diarrhoea. Harrison describes some variegated forms.

Ref.: (Bircher); (Chittenden); Eliovson; Encke; (Graf); Hamlin; Harrison; Kunkel 69; (Menninger 62, 64); Neal; Uphof ().

Fig. 99. *Metrosideros excelsa*; natural size

Psidium cattleianum Sabine – Strawberry Guava

Small tree up to 6 (8) meters tall; trunk slender, soon ramified. Bark grey and scaly. Crown small, dense and dark. Evergreen. Leaves opposite, decussate, short-stalked, subcoriaceous, dark green and shiny above; blade obovate, cuneate at base, 4 to 6 cm long and 2 to 3 cm wide; venation rather inconspicuous. Flowers axillary, solitary, with white stamens about 1 cm long. Fruits subglobose, 2 to 2,5 cm in diameter, juicy, with a brownish or purplish skin and a persistant calyx; edible, and tasting rather like strawbarries.

The Strawberry Guava is native in South America (Brazil to Uruguay) and is sometimes cited as *Psidium littorale* Raddi; by Chittenden both names are cited, somewhat doubtfully, as two different species. – Much cultivated in tropical and subtropical countries; fruits are eaten raw or in jellies; the juice is recommended for refreshing drinks. Propagation from seeds. Its var. *lucidum* seems to have larger leaves and bigger, yellowish fruits. – Alternative names: Gooseberry Guava (Irvine), Purple Guava, and Cattley Guava.

Ref.: Adams; Bircher; Burkill; Calabria; Chittenden; Cobley; Encke; Graf; Irvine; Kennard & Winters; Kunkel 69; (Lombardo 61); Menninger 64; (Molesworth Allen); Neal; (Purseglove); Uphof.

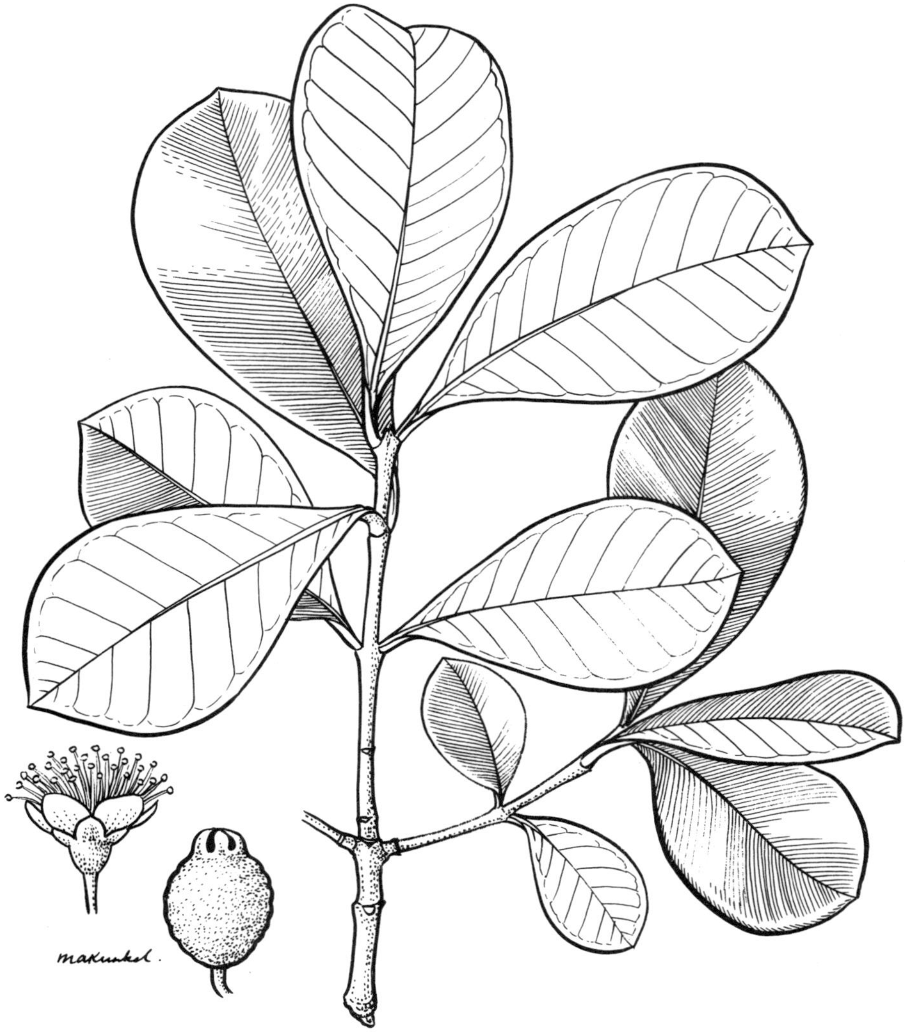

Fig. 100. *Psidium cattleianum*; slightly reduced

Psidium guajava Linné – Common Guava

Small, open-branched tree up to 10 m in height; trunk heavy. Bark scaly, chestnut-coloured. Young branches angular; foliage deciduous. Leaves short-petioled, bright green, subcoriaceous, sometimes folded or slightly undulate, base truncate; 8 to 12 cm long and 4 to 6 cm wide; strong-veined. Flowers solitary, axillary, stalked, with long white stamens; fully open flowers up to 3 cm across. Fruits very variable in size according to the particular variety; up to 7 cm in diameter, with a yellow skin and a persistant calyx. Pulp usually pinkish, with many small pips; fruits eaten raw or made into jelly or the famous 'guava cheese'. Stored fruits may give off a musky smell.

The Common Guava (or just Guava) is native in the region from the West Indies down to Peru. It is a widely cultivated species of which several varieties are known. Leaf-buds are used in Hawaii for an astringent tea (Neal). The bark is rich in tannin. According to Barrett the raw fruit is very rich in vitamin C and is said to contain 2 to 5 times that of fresh oranges. The wood is used in carpentry. For detailed information on the history of cultivation and general uses see the works by Burkill and Purseglove.

Propagation from seeds and suckers. Best grown in a somewhat cooler climate at lower altitudes. The species is naturalized in West Africa (Irvine) and other places where it can form dense thickets near to the sea.

Ref.: Adams; Barrett 56; Bircher; Burkill; Calabria; Chittenden; Cobley; Corner; Encke; Esdorn & Pirson; Graf; Irvine; Kennard & Winters; Little & Wadsworth; Lombardo 61; Long & Lakela; Molesworth Allen; Neal; Pesman; Purseglove; Soukup; Uphof.

Fig. 101. *Psidium guajava*; flowering branch and details = 3/5

Syzygium cuminii (L.) Skeels – Java Plum

A fruit tree native in Southeast Asia often also cited as *Syzygium* (or *Eugenia*) *jambolana*. Alternative common name: Jambolan. The species is much cultivated for its fruits in several tropical and subtropical countries; in the Canary Islands grown as windbreak, shade tree and for ornamental purposes. Small to medium-sized tree up to 20 m tall; trunk high, 40 to 50 cm in diameter, with a rough, rather warty grey bark. Crown spreading; foliage evergreen. Branches somewhat knotty; branchlets very pale. Leaves opposite, stalked, pendent, coriaceous, glossy green; blade lanceolate, up to 18 cm long and 5 cm wide; with numerous fine lateral veins. Flowers small, white, in subterminal, much branched cymes. The edible berries are roughly plum- or pear-shaped, up to 2,5 cm long, lilac-purple or nearly black, and each contain one single nut-like seed.

The fruits may be eaten raw but are very acid; they are best used in jams and are also made into 'a wine with a very fine 'bouquet'' (Bircher). Burkill reports a vinegar made from the fruits in India. The bark is used for dyeing, tanning, and home remedies; the leaves are said to be used to feed silkworms. According to Uphof there are some seedless varieties; the same author states that the 'seeds are astringent, diurtetic (sic!); claimed to be useful in diabetes; it reduces sugar in urine in brief time'. For further applications see Burkill. – The Java Plum or Jambolan is resistant to wind and recommended for seaside gardens. Propagation from seeds and cuttings.

Ref.:Adams; Barrett 56; Bircher; (Burkill); Calabria; (Chittenden); (Corner); Encke; Irvine; Kennard & Winters; Kunkel 69; Long & Lakela; Menninger 59; (Molesworth Allen); (Neal); (Uphof).

Fig. 102. *Syzygium cuminii*; flowering branch and leaf = 3/5, fruit = almost natural size

Syzygium jambos (L.) Alston – Rose Apple

Often cited as *Eugenia jambos* L., this fruit tree seems to be native in Southern or Southeastern Asia, and is now widely cultivated especially in subtropical countries. It makes an effective shade tree and is also admired as an ornamental. Its fruits are eaten raw or used in sauces and jellies.

Tree 8 to 12 m in height; trunk short, reaching over 60 cm in diameter but often branching from the base. Bark grey-brown, finely fissured. Young branches reddish. Crown spreading; foliage evergreen. Leaves opposite and stalked, appearing to be in 'rows'. Blade lanceolate, subcoriaceous, shiny, usually 10 to 15 cm long and 2,5 to 3 cm wide; nerves prominent. Cream-coloured flowers with long stamens clustered in very showy false umbels; individual flower-heads up to 8 cm across. Fruits fragrant (rose-scented), globose berries 2,5 to 4 cm in diameter, with a yellowish, later brownish skin; calyx persistant. Seeds 1 to 3.

Propagation from seeds. Wood used for constructions, and the bark for tanning and dyeing. All bark, roots, fruits and seeds are used in home medicine (Barrett, Burkill). Any excess in application however may have poisonous effects (Burkill).

Ref.: Adams; (Barrett 56); Bircher; (Burkill); (Chittenden); (Corner); (Eliovson); Encke; (Kennard & Winters); Kunkel 69; (Little & Wadsworth); Long & Lakela; Menninger 59; (Molesworth Allen); (Neal); (Purseglove); (Uphof).

A third species of this genus – the Clove tree, *Syzygium aromaticum* (L.) Merr. & Perr. (*Eugenia caryophyllus*) from the Moluccas – is also cultivated but extremely rare in these islands. Its dried flower-buds are the source of the commercial Clove and are used as spice.

Fig. 103. *Syzygium jambos*; drawings = 3/5

ANACARDIACEAE, the Mango family

This family of almost 60 genera and some 550 to 600 species is widely distributed in the tropics. One species of Sumach (*Rhus coriaria*) and two species of Pistachio (*Pistacia atlantica, P.lentiscus*) are native in these islands; according to modern taxonomists the latter genus now forms an independent family: Pistaciaceae.

Mangifera indica Linné – Mango

A tree of South Asian origin and widely cultivated throughout the tropics of both hemispheres. According to Neal there are more than 500 varieties known in India alone, and recently created improved cultivars might, one expects, be distributed from tropical North and Central America where the Mango is highly appreciated as a fruit and shade tree. The Mango is a sacred tree to Buddhists and Hindus, and seems to have been in cultivation for over five thousand years (Barrett).

An evergreen tree up to 20 (30!) m tall, this species has a dense, broad crown and a short, stout trunk which may reach 80 to 100 cm in diameter. Bark brown and fissured or, when older, scaly. Freshly cut twigs exude a whitish latex. The gum and bark is used locally as an astringent. The timber is said to be non-durable but is applied for planking, doors, packing-cases etc. Although in India found growing up to over 1.000 m above sea level, the tree does best in coastal zones.

The leaves are lanceolate, subcoriaceous, dark green, up to 30 cm long and 5 to 7 cm broad; they are petiolate, often slightly curved or bent downwards. Young leaves can be reddish and glossy. Flowers very small, yellowish or whitish, somewhat fragrant, in large panicles. The large and leathery skinned aromatic fruits may reach up to 15 (30) cm length, are juicy-fibrous in primitive varieties whereas the improved cultivars have truly fleshy fruits with a juicy pulp tasting slightly of turpentine. The Mango is eaten raw or makes an excellent jelly or jam; unripe fruits are the base for Mango Chutney, and are also said to have diuretic properties.

According to Burkill and Uphof a yellow dye is obtained from the urine of cows which have been fed with Mango Leaves; this dye is used in the preparation of water and oil paints and is reputed to be resistant to light. The seeds are medicinal and act as a vermifuge. Leaves are astringent. According

Fig. 104. *Mangifera indica*; branch and fruit = 3/5, solitary flower = × 1,5

to Irvine 'stewed mango leaves and buds are taken internally in Nigeria for fever, while the smoke from mango wood causes dermatitis'. – Although plants are usually grown from seeds, grafting (with special variety stock) is recommended in order to obtain low growing, earlier and rich fruiting trees.

Ref.: On varieties and uses see Bircher, Burkill, Cobley, Corner, Esdorn & Pirson, Molesworth Allen and, especially, Purseglove (1: 24–32); the latter author's treatment makes really good reading. Many references are given by Mortensen & Bullard. An interesting account on varieties can be found in the work by Calabria where both – 'mangos and mangas' – are cited. For further references see Barrett 56; Blackwell & Dodson, Chittenden, Encke, Graf, Irvine, Keay & al., Little & Wadsworth, Martinez, Neal, Pesman, and Uphof.

Schinus molle Linné – Peruvian Pepper Tree

Although there are about 30 species of *Schinus* (native from Mexico to Chile and Argentina), none of them have reached the popularity of the Peruvian Mastic or Pepper Tree as commonly grown in these islands. The name 'Brazil Pepper-tree' given by Uphof may have been confused with the second species described further on, and the name 'California Pepper tree' given by several authors may be because of the frequent appearance of this species in California, as a tree of parks and gardens, roadsides and even on slopes or bordering fields; in any case *Schinus molle* is a native of Sub-andean South America. According to Neal it is rather unpopular in Hawaii because it harbours the so-called 'black-scale', a pest of *Citrus* trees. In the Canary Islands it harbours a kind of white scale or woolly louse, much visited by ants and therefore highly unpopular in plant nurseries.

It may harbour a pest, or several pests, but who would like to miss this gracious tree which is little demanding and which gives one the illusion of being surrounded by 'weeping willows'? The Peruvian Pepper Tree may reach 10 to 12 m in height and has a short trunk with fissured or scaly bark of a light greyish-brown colour. It has a wide-branching crown with slender arching or hanging branches which, when cut, exude a whitish resin or gum used in South America as chewing gum ('American mastic'). The powdered bark is given as a purgative for domestic animals (Uphof), and both bark and leaves are used in America (Burkill) to combat swellings and sores and even in the treatment of veneral diseases. However, according to the latter author, the pollen of this species seems to produce a kind of irritation related to the so-called 'hay fever'.

Foliage evergreen. Leaves pinnate, up to 25 cm long, composed of 12 to 15 pairs of opposite to alternate narrow leaflets which are quite fragrant (peppery) when crushed. Flowers very small, greenish-white, in lax, hanging inflorescences (male and female separate). Female trees may produce rose-coloured, spherical fruits which are single-seeded and very decorative when cut in bunches. In some countries the dried seeds are used for adulterating pepper (Neal). According to the same author, in Mexico the fruits are collected to produce a beverage, and an intoxicant is made from them when

Fig. 105. *Schinus molle*; flowering branch and fruits = 1/2, solitary flower = × 3

fermented. The Peruvian Indians also use the fruits to produce a fermented drink called 'chicha de molle' (Soukup).
In parks especially, this species is almost irreplacable as it asks for little attention and its decorative value makes up for any known disadvantages. It grows easily from seeds and sizeable specimens may be obtained after 5 to 6 years. – Alternative names: Molle, Weeping Pepper, and Peruvian Mastic.

Ref.: Bircher; Blackwell & Dodson; Burkill; Chanes; Chittenden; Degener; El Hadidi & Boulos; Encke; Graf; Harrison; Kunkel 69; Lombardo 58, 64; Moeller 68; Neal; Pesman; Schaeffer; Soukup; Tutin 68; Uphof.

Schinus terebinthifolius Raddi – Brazilian Pepper Tree

This species is a native of the Chaco region (S.Brazil, Paraguay, N.Argentina). Also known as Aroeira, or Christmas Berry Tree, the species thrives well in coastal gardens where it may reach 8 to 10 m height. Its short trunk is often curved, of greyish colour and is furrowed or scaly. Branches not pendulous. The tree is resinous and once produced the so-called 'Bálsamo de Misiones'. The persistant leaves are imparipinnate; with 7 to 9 oval-shaped or obovate leaflets up to 5 cm long, and used as a tonic and astringent. Flowers small and greenish, in paniculate inflorescences. Fruits globose, glossy, bright-scarlet, in dense bunches; fruiting branches are often cut and may be kept a long time in a flower-vase.
The Brazilian Pepper Tree is easily grown from seeds and serves for reafforestation in warmer climates. However, according to Neal in Hawaii it grows like a weed due to seeds distributed by birds. Little & al. mention that propagation is also possible from cuttings. According to the same authors 'it is reported that this species causes a skin irritation in some persons'. – Several varieties seem to be known. The species is naturalized in Southwestern Europe (Tutin).
Occasionally a third species – *Schinus* cf. *weinmanniifolius* Engl. – is also found in gardens but it has always remained a rare plant in the Canary Islands.

Ref.: Adams; Barrett 56; Bircher; Blackwell & Dodson; Chittenden; Degener; El Hadidi & Boulos; Encke; Harrison; Kunkel 69; Little & al.; Menninger 64; Neal; Schaeffer; Tutin 68; Uphof.

Other important species of Anacardiaceae but rarely seen in Canary gardens include the Cashew Nut (*Anacardium occidentale* L.), a tree of spreading habit native in tropical America and cultivated for its edible, kidney-shaped seeds or nuts; species of *Spondias* including the Hog or Spanish Plum (also native in tropical America and widely cultivated in Africa and Asia), and *Cotinus coggygria* Scop. (*Rhus cotinus*), the Smoke Tree from Southern Europe and Asia (towards China). *Rhus typhina* L., the Staghorn or Lemonade Tree from eastern North America is popular in European gardens.

Fig. 106. *Schinus terebinthifolius*; drawings = 3/5 from: Kunkel 'Arboles exóticos'

PISTACIACEAE, the Pistachio family

As mentioned previously this family has been segregated from the Sumach family and seems to consist of one single genus, with 10 or 12 species of which *Pistacia atlantica* and *P.lentiscus* are Mediterranean elements native also in the Canary Islands. *Pistacia vera* L., from the Middle East and source of the Pistachio nut, has been introduced but seems to have encountered difficulties in cultivation. The remaining exotic is

Pistacia chinensis Bunge – Chinese Pistachio

A sizeable tree up to 15 (20) meters tall with a high trunk and a fissured or furrowed bark. The species is obviously deciduous but some specimens change their leaves almost unnoticed. The leaves are pinnate and leaflets may be opposite or alternate; leaflets ovate-lanceolate, up to 12 cm long, subcoriaceous and dark green; they are unequal at the base and have a tapering apex. Shoots and leaves are somewhat downy when young, and the leaves take a kind of crimson colour towards the autumn.
Flowers small and pale green, in dense conical inflorescences up to 15 cm long. Fruits globose-obovoid, 1,5 to 2 cm in diameter, reddish-brown but turning bluish when ripe.

According to Uphof the wood of this tree is used for rudderposts, and young shoots are eaten as a vegetable. As reported by Neal and other authors, the species is grown as a shade tree and used as stock for the true Pistachio Nut tree.

Ref.: Chittenden; Harrison; Neal; Uphof.

Fig. 107. *Pistacia chinensis*; all drawings = 1/2 except flower detail (= × 2)

SIMAROUBACEAE, the Ailanthus family

Although this family of tropical and subtropical distribution seems to consist of about 20 genera with over 150 species in total, only very few species are widely cultivated, and one of these known as

Ailanthus altissima (Mill.) Swingle – Tree of Heaven

has established itself in many subtropical countries. In fact, the Tree of Heaven, from China, has become a 'weed' of Mediterranean and American countries and is on the way to become rather troublesome on moister slopes of some of the Canary Islands.

This small tree up to 10 or 15 meters in height propagates from seeds but worse than this is its habit of spreading from root-shoots and suckers soon covering areas of considerable extension. These young shoots are hardly woody but have brittle stems of very vigorous growth. The species is deciduous. Leaves pinnate up to 80 cm long, stalked, with 9 to 25 leaflets which are herbaceous and softly downy; leaflets oblong to oblong-lanceolate, irregularly toothed or even slightly incised; base unequal with glands at edges of lower lobes. Leaves and flowers have an unpleasant smell. Flowers small, white or whitish-green, in large terminal panicles up to 40 cm long. Fruits long, flat, winged, with a single seed in the middle.

According to Ruiz de la Torre the wood is yellowish or slightly rose-coloured, provides a reasonable charcoal and might be used to manufacture cellulose. Old specimens provide a hard wood difficult to split but easy to work and to polish (Uphof; Hegi). Burkill reports that roots, bark and leaves have medicinal properties (dysentery, haemorrhage, and urinary difficulties) but should be applied with care as an overdose may result in poisoning. – As for gardening purposes the species could well be considered as 'graceful' it it were not for its habit of invading the surrounding countryside. Hegi reports 27 m as the maximum height of this species, and Bean cites a specimen from Devon, 95 feet high (over 30 meters). – The species is often wrongly cited as *Ailanthus glandulosa* Desf.

Ref.: Bean; Bircher; (Burkill); Chanes; Chittenden; Encke; Fitschen; Harrison; (Hegi V/1); Lombardo 58; Long & Lakela; Menninger 64; Mitchell; Moeller 68; Neal; Polunin & Smythies; Ruiz de la Torre; Tutin 68; (Uphof).

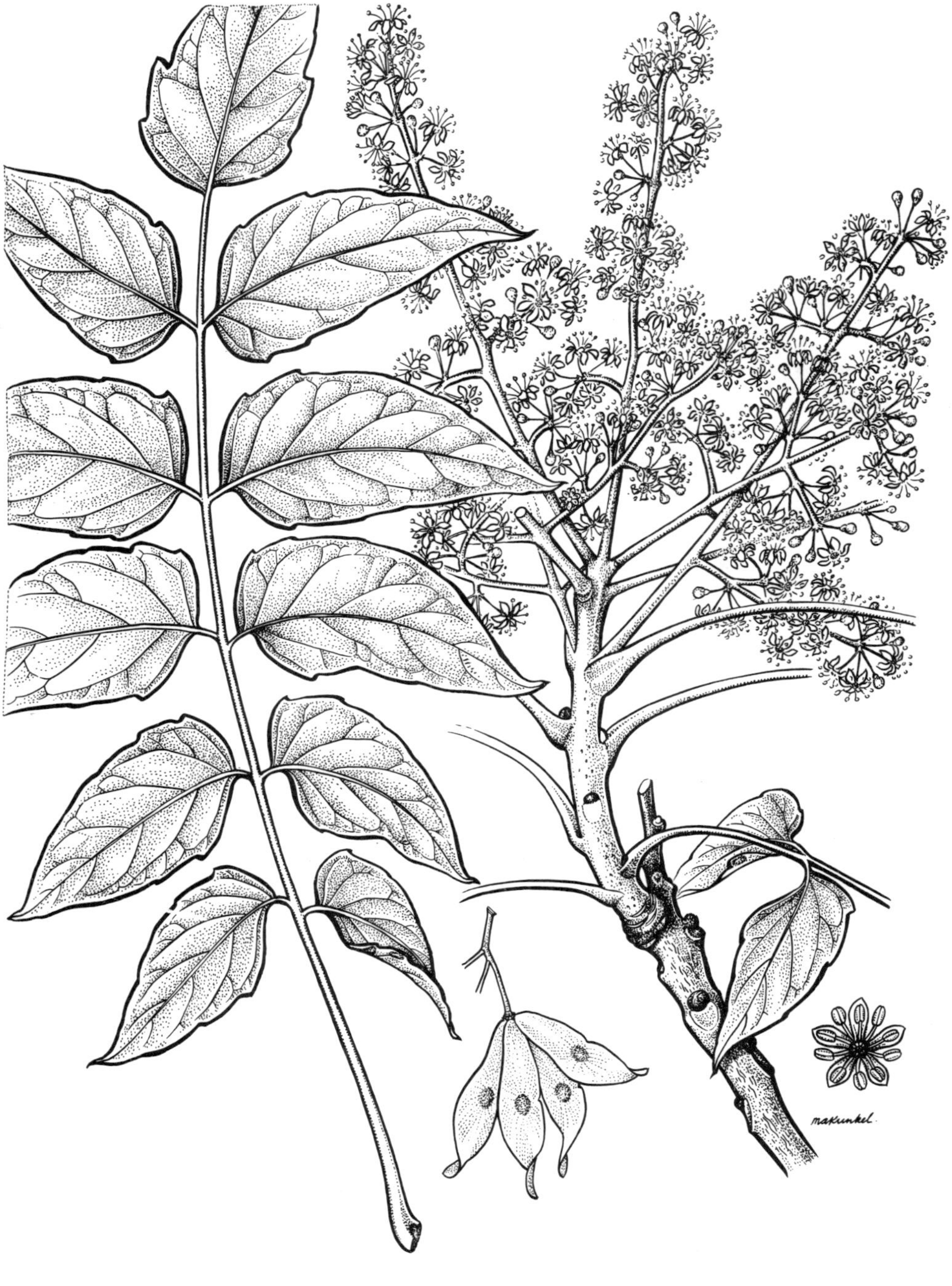

Fig. 108. *Ailanthus altissima*; leaf = 1/3, flowering branch and fruits = 1/2, flower detail = 1,5

RUTACEAE, the Citrus family

A large family of about 150 genera and 900 species (Willis), according to Purseglove 130 genera and 1.500 species; mostly trees and shrubs native in the Southern Hemisphere (S.Africa, Australia). Many ornamentals (*Diosma*, *Fortunella*, *Murraya* etc.), important fruit trees (*Citrus*) and some medicinal plants (*Ruta*). Three shrubby species of *Ruta* are endemic in the Canary Islands.

Casimiroa edulis La Llave & Lex. – White Sapote

Fruit tree from Central America, now widely grown in many subtropical regions. About 9 to 12 m tall; trunk short and up to 50 cm in diameter. Bark dark grey, smooth or finely fissured, with lenticels. Crown spreading; foliage apparently evergreen. Leaves alternate, long-stalked, with 3 to 5 leaflets; petiole up to 12 cm long. Leaflets ovate or oblong-lanceolate, subcoriaceous, glossy green, up to 15 cm long; margin entire, slightly crenate or somewhat undulate. Flowers small, greenish-white, in axillary clusters. Fruits globose and fleshy, thin-skinned, yellowish when ripe, up to 7 cm in diameter, their shape and colour depending on the particular variety; pulp yellow, sweet, edible. Seeds 3 to 5, rather large.

Propagation from seeds and cuttings. A fast growing species best cultivated in somewhat cooler zones at lower altitudes. Fruits eaten in desserts. According to Martinez the seeds are sedative. Otherwise the White Sapote makes a good shade tree and is also quite decorative.

Ref.: Barrett 56; Bircher; Burkill; Chittenden; Kennard & Winters; Kunkel 69; Martinez; Mortensen & Bullard; Neal; Pesman; Uphof.

Fig. 109. *Casimiroa edulis*; drawing = 1/2

Citrus spp. – Citrus fruit trees

The genus *Citrus* consists of 11 (Neal), 12 (Willis) or 16 species (Purseglove, Synge, Uphof), of *many* varieties, forms and cultivars besides numerous hybrids including an intergeneric hybrid with *Poncirus*. These 11 or 16 species have been given many names; the Catalogue of *Citrus* species and cultivars including synonyms, published in 1969 by Carpenter & Reece, occupies over 60 pages.
The genus is native in East and Southeast Asia. *Citrus* is known to have been in cultivation for over three thousand years, and the most important species are so well known that little remains to be said about them. The different species and cultivars grown in our gardens and plantations have certain characters in common: they are small evergreen trees with undivided shiny leaves, very fragrant white flowers, and globose, juicy edible fruits which, technically speaking, are berries. *Citrus* are grown from seeds and basal shoots both of which should be grafted by budding with the species or variety desired. They may be grafted onto the same species or onto others (viz orange on mandarins, or viceversa); ingenious gardeners who possess only one specimen of *Citrus* may graft several species or varieties on the one common stock available. Most species thrive best in subtropical climates but are also seen in the humid tropics. For their cultivation, protection and diseases see Purseglove (pp. 495–522). The following enumeration deals with the more common species only, without detailed notes on cultivars.

Citrus aurantifolia (Christm.) Swingle – Lime

Leaves dark green, relatively small, ovate-elliptic, up to 8 cm long and 5 cm wide; margin entire, petiole narrowly winged. Spines short. Fruit more or less spherical, 4 to 6 cm in diameter, greenish-yellow, skin thin; pulp greenish and very acid. – Common hybrids: Mandarin-lime (with *C.reticulata*), Sweet Lime (with *C.medica*), and Limelo (with *C.paradisi*).

Citrus aurantium Linné – Sour or Seville Orange

Leaves pale green, ovate or elliptic, 5 to 15 cm long and up to 7 cm wide; margin more or less crenate, petiole broadly winged. Spines either present or absent. Fruit almost globose, up to 8 cm in diameter, orange-red, thick-skinned; pulp bitter, very sour. – Suspected hybrid: Bittersweet Orange (with *C.sinensis*).

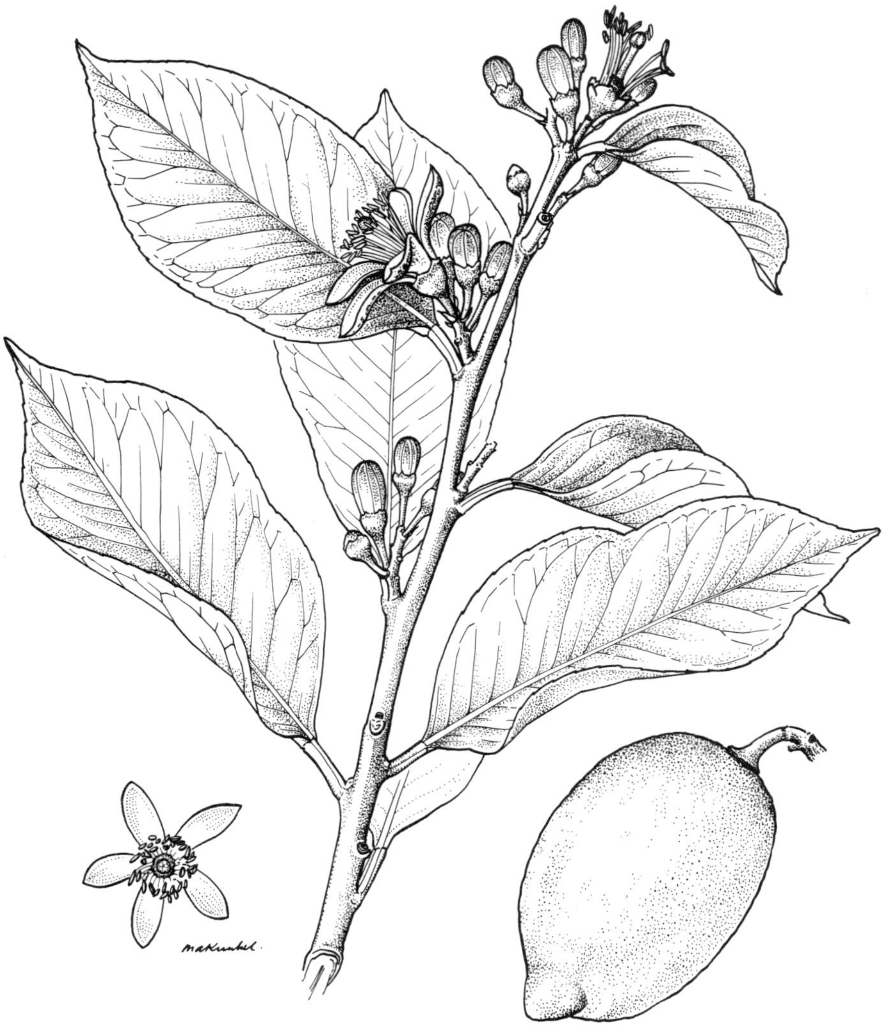

Fig. 110. *Citrus limon*; drawing about 1/2

Citrus grandis (L.) Osbeck – Pomelo, or Shaddock

Leaves dull green, often pubescent on the midrib of the undersurface; elliptic in outline, up to 20 cm long and 12 cm wide; margin entire or slightly crenate, petiole broadly winged. Spines present or absent. Fruit globose or pear-shaped, up to 30 cm in diameter; skin thick, yellowish; pulp pale yellow or pinkish, fairly sweet.

Citrus limon (L.) Burm.f. – Lemon

Young branches often reddish. Leaves more or less ovate, up to 12 cm long and 6 cm wide; margin serrate or slightly sinuate, petiole usually not winged. Spines, if present, stiff and up to 2 cm long. Fruit ellipsoid, 8 to 10 cm long, with a terminal knob or nipple; skin relatively thin, rough, light yellow; pulp pale yellow, very sour. – The real Rough-skinned Lemon is probably a hybrid with *C.medica.*

Citrus medica Linné – Citron

Leaves oblong-elliptic, up to 20 cm long and almost 10 cm wide; margin shallowly toothed, petiole short and usually not winged. Spines stout, up to 5 cm long. Fruit oblong-ovoid, up to 20 cm long, with a very thick, yellow, sometimes wrinkled skin; pulp greenish, sour, rather dry. – Hybrids: Sweet Lime (with *C.aurantifolia*), and probably also with *C.limon.*

Citrus paradisi Macfayden – Grapefruit

Leaves pale green, ovate-elliptic, up to 12 cm long and 8 cm wide; margin often crenulate, petiole usually broad-winged. Spines short, or absent. Fruit globose, very large, up to 15 cm in diameter, with a yellow, rather thin skin; pulp yellow or pinkish, very juicy. – Hybrids: Tangelo (with *C.reticulata*), Limelo (with *C.aurantifolia*), and Chironja (with *C.sinensis*).

Citrus reticulata Blanco – Mandarin and Tangerine

Leaves lanceolate, narrow-oblong, ovate or almost elliptic (many varieties), up to 8 cm long; margin crenate or somewhat undulate, petiole narrowly winged or wingless. Spines small or absent. Fruit nearly spherical or slightly flattened, up to 8 cm in diameter; skin thin, rather loose, yellow to orange-red; pulp orangy, sweet and juicy. Some varieties are seedless. Fruits of the true Mandarins have a more yellowish skin whereas the Tangerines ('Clementine', 'Dancy') are orange-red. – Frequent hybrids: Mandarin-lime (with *C.aurantifolia*), Tangelo (with *C.paradisi*), and Tangor (with *C.sinensis*).

Citrus sinensis (L.) Osbeck – Sweet Orange

Twigs ridged or angled. Leaves ovate or oblong-elliptic, up to 15 cm long and 7 cm wide; margin entire or slightly serrate, petiole narrowly winged. Short stout spines, especially on younger specimens. Fruit more or less spherical, up to 8 (12) cm in diameter, skin thin or medium thick (varieties!); pulp orangy, juicy, sweet or semi-acid. – Known hybrids: Tangor (with *C.reticulata*), Bittersweet Orange (with *C.aurantium*), and Chironja (with *C.paradisi*).

All species cited above are famous for their fruits which are eaten raw, in salads, or made into preserves and juices. All fruits are rich in vitamins, most have medicinal properties (Burkill, Uphof), and all are hopelessly overloaded with synonyms adding to the often confusing names of known cultivars. *Citrus* trees are sometimes attacked by a certain fruit-fly. In the Canary Islands production decreased considerably after the arrival of the tiny 'white fly', with subsequent problems of ants and black fungi.

According to Esdorn & Pirson in 1970 the world harvest of Mandarines and Sweet Oranges was estimated at 30.252.000 tons of which 8.637.000 t were produced within the territory of the United States (Spain: 2.150.000 t); however Spain seems to be the most important export nation (1.029.300 t, against USA = 528.000 t).

Ref.: Adams; Barrett 56; Bircher; Burkill; Calabria; Carpenter & Reece; Chanes; Chittenden; Cobley; Corner; Eliovson; Encke; Esdorn & Pirson; Font Quer, Graf; Irvine; Little & al.; Long & Lakela; Menninger 64; Molesworth Allen; Mortensen & Bullard; Neal; Perry; Pesman; Polunin & Huxley; Purseglove; (Rauh); Schütt; Soukup; Synge; Uphof.

MELIACEAE, the Mahogany family

The Mahogany family includes many genera and numerous species of trees and shrubs mainly natives of warm regions. According to Neal (1965) and Willis (1973) the family consists of 50 genera and 1.400 species; Chittenden (1965) cites 'about 40 genera with nearly 700 species', whereas a recent monographic study by Pennington & Styles (1975) restricts the family to 51 genera, with approximately 550 species. However, as we are dealing here with a few species only, we should remain little disturbed by the remarkable differences in the specific numbers as given above.

Azadirachta indica Juss. – Margosa, or Neem Tree

Closely related to *Melia* and often included in the genus (as *Melia azadirachta* L., or *M.indica* Brand.), this tree native in Southern and Southeastern Asia differs from *Melia* by certain pronounced morphological and anatomical characters as explained by Pennington & Styles.

An evergreen tree up to 15 m tall; bark grey and scaly, trunk short, up to 60 cm in diameter. Crown with upright or spreading branches and dark green, pinnate leaves up to 30 cm long. Leaflets 7 to 17, opposite or almost so, somewhat curved, oblong-lanceolate and up to 7 cm long, with a noticeable onion smell. Margin strongly serrate; base cuneate and unequal. Flowers white and fragrant, in large panicles. Fruits berry-like, oval-oblong, yellow to reddish-brown and only slightly fleshy; single-seeded.

The timber is workable but unpleasant smelling while fresh. Rootbark is considered a useful astringent and febrifuge, and it is said that the entire tree has medicinal properties of one sort or another. The antiseptic resin is added to toothpaste, soap and lotions (Neal). Decaying leaves are supposed to kill injurious animals; the margosa oil (from seeds) remains moist for long periods (a non-drying oil) and is used to cure skin diseases, and leaves kept in libraries and herbaria 'keep out book-mites and other insects' (Uphof). Burkill who dedicated over two pages to the use and properties of this species reports anti-malarial and other virtues, and mentions that 'bark and young fruits are tonic and anti-periodic' besides which the flowers seem to be a stimulant, tonic and stomachic. The tree is grown from seeds and is relatively fast growing. It is popular in gardens and often seen as a roadside tree.

Ref.: (Chittenden); (Corner); (Burkill); Irvine; Keay & al.; Kunkel 69; Little & al.; Neal; Pennington & Styles; Uphof.

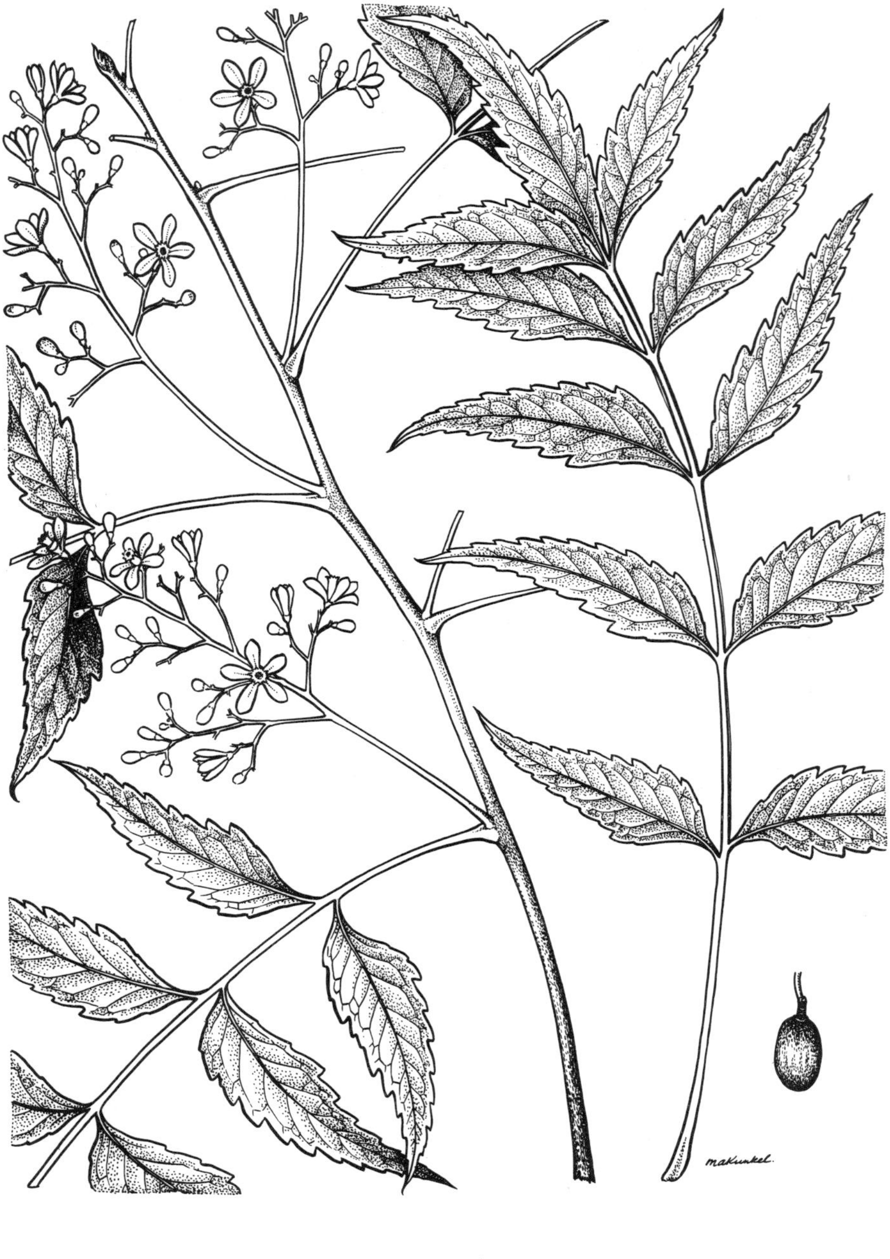

Fig. 111. *Azadirachta indica*; drawings = 1/2

Melia azedarach Linné – Persian Lilac

This very decorative small to medium-sized tree supposed to be a native of Southern Asia (from Iran to India and South China) has been designated many common names: Persian lilac, Indian lilac, West Indian lilac, Lilac, China tree, Hoop tree, Umbrella tree, Pride of India, Pride of China, Paradise tree, Chinaberry tree, Bead tree etc. It seems to be widely cultivated in warmer countries, and it is appreciated as a shade tree as well as for its showy flowers. The species seems to be sufficiently naturalized in Europe to be dealt with in 'Flora Europaea'.

Growing up to 12 or 15 m in height, the Persian Lilac has a pronounced trunk 40 to 50 cm in diameter, with a fissured, dark grey or brownish bark. Branches spreading. Deciduous. Leaves alternate, petiolate, bipinnate, almost triangular in outline and up to 60 cm long. Leaflets numerous, imparipinnate, herbaceous, oblong or ovate; they may reach up to 12 cm in length. Margin slightly serrate or, on young shoots, coarsely toothed. Branched flower clusters in loose terminal panicles up to 30 cm long. Flowers can most aptly be described as lilac-coloured, the petals being pale and the staminal column quite dark; very fragrant, on slender stalks. Fruit (a drupe) yellowish and, when ripe, slightly wrinkled, up to 2 by 1 cm, with 3 to more seeds; the decorative clusters of ripening fruits make this an attractive tree even when the leaves have fallen.

The wood is used as fuel, and the reddish-brown heartwood in cabinet making, for tool handles, cigar boxes etc. Bark of roots, according to Uphof, is used medicinally and has anthelmintic properties. Leaves have vermifuge and diuretic applications; when depostied in footwear they are supposed to keep insects away (Neal). Burkill reports that 'poultices of leaves and flowers are used in India for nervous headaches'. An extract made from the flowers acts as a deterrant against lice, and the pulverized bark as a remedy for leprosy and other skin diseases. The fruits are poisonous to man and most animals except birds, and have insecticidal properties; oil from the seeds is found to be suitable for oil lamps, and the seeds themselves are used to make rosaries and beads.

This species regenerates freely from seeds and is considered a fast-growing tree. However, as the wood is somewhat brittle it is frequently damaged by strong winds. The Texas Umbrella Tree (var. *umbraculiformis*) is cited by Eliovson who gives our typical variety (var. *azedarach*) still another common name: South African Syringa.

Ref.: Adams; Barrett 56; Beyron; Bircher; Burkill; Chittenden; Corner; Degener; El Hadidi & Boulos; Eliovson; Harrison; Irvine; Keay & al.; Kunkel 69; Little & Wadsworth; Lombardo 58; Long & Lakela; Martinez; Menninger 62, 64; Moeller 68; Neal; Pesman; Polunin & Huxley; Polunin & Smythies; Schaeffer; Tutin 68; Uphof.

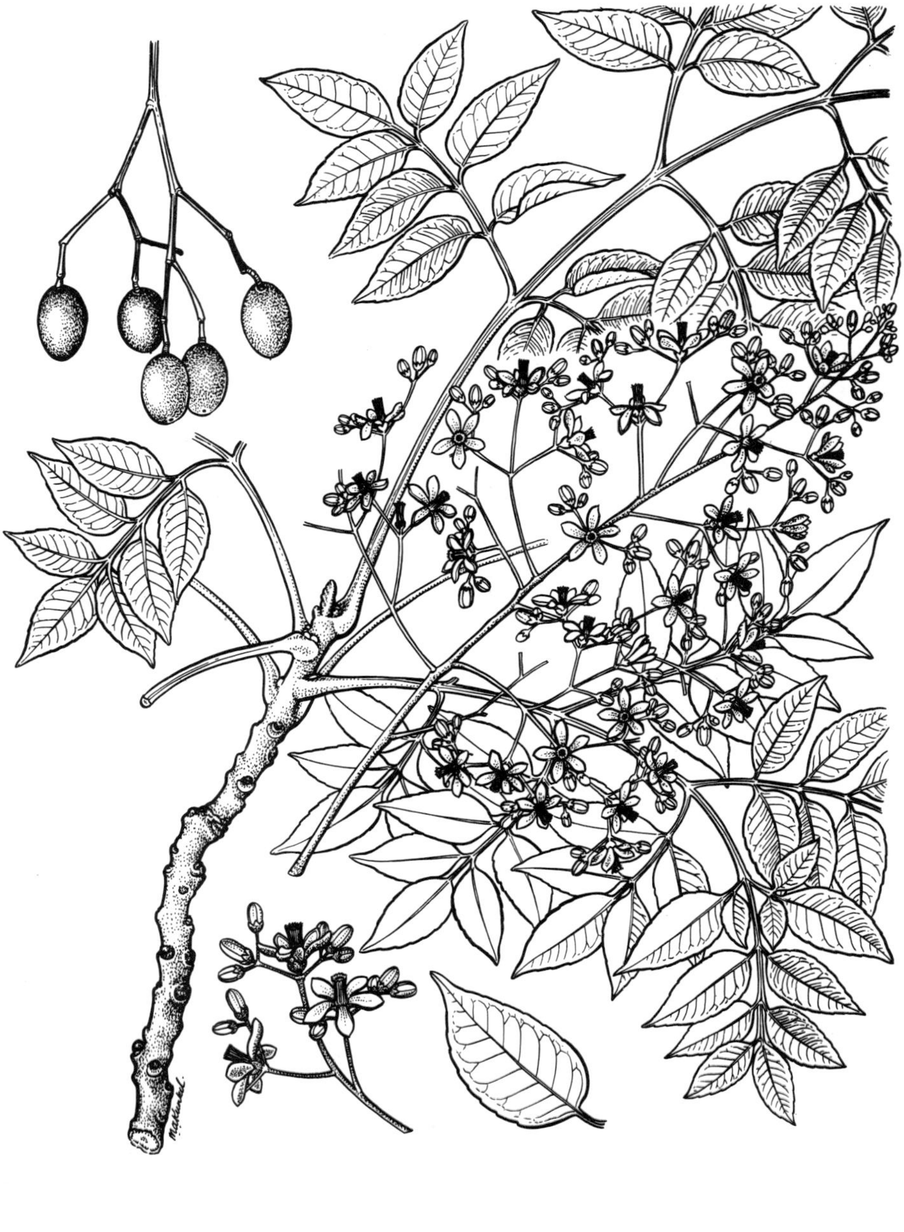

Fig. 112. *Melia azedarach*; all drawings = 1/2

Meliaceae

The Mahogany family has perhaps produced the most highly appreciated timber trees known. Genera like *Cedrela*, *Entandrophragma*, *Guarea*, *Khaya*, *Lovoa*, *Swietenia*, *Toona*, *Trichilia* etc. are classed as very valuable species, especially for cabinet making and other furniture. Some of these species are now endangered because of ruthless exploitation, and only few are grown commercially in plantations. They are seldom found in the Canary Islands, i.e. *Khaya senegalensis*, *Swietenia mahagoni*, *Toona sinensis* and *Cedrela odorata* have been tried of which only

Cedrela odorata L., – West Indian Cedar

a 'Cigar Box tree' from the West Indies and northern South America has proved successful in some gardens. In its native environment reaching over 30 meters in height, this species grows in some local gardens as a twisted tree up to 10 m, with a furrowed grey or brownish bark. The foliage is briefly deciduous. Leaves paripinnate up to 50 cm long, with 8 to 12 pairs of unevenly shaped, oblong-lanceolate leaflets. Flowers whitish-green, in hanging panicles. Woody seed-capsules 1 to 1,5 cm in diameter with brownish, winged seeds. The tree is appreciated because of its valuable timber (furniture, cabinet-making, cigar-boxes etc.), and has limited medicinal properties.

Ref.: Adams; Barrett 56; Bircher; Burkill; Chittenden; (Irvine); Keay & al.; Kunkel 69; Little & Wadsworth; Neal; Smith; Uphof.

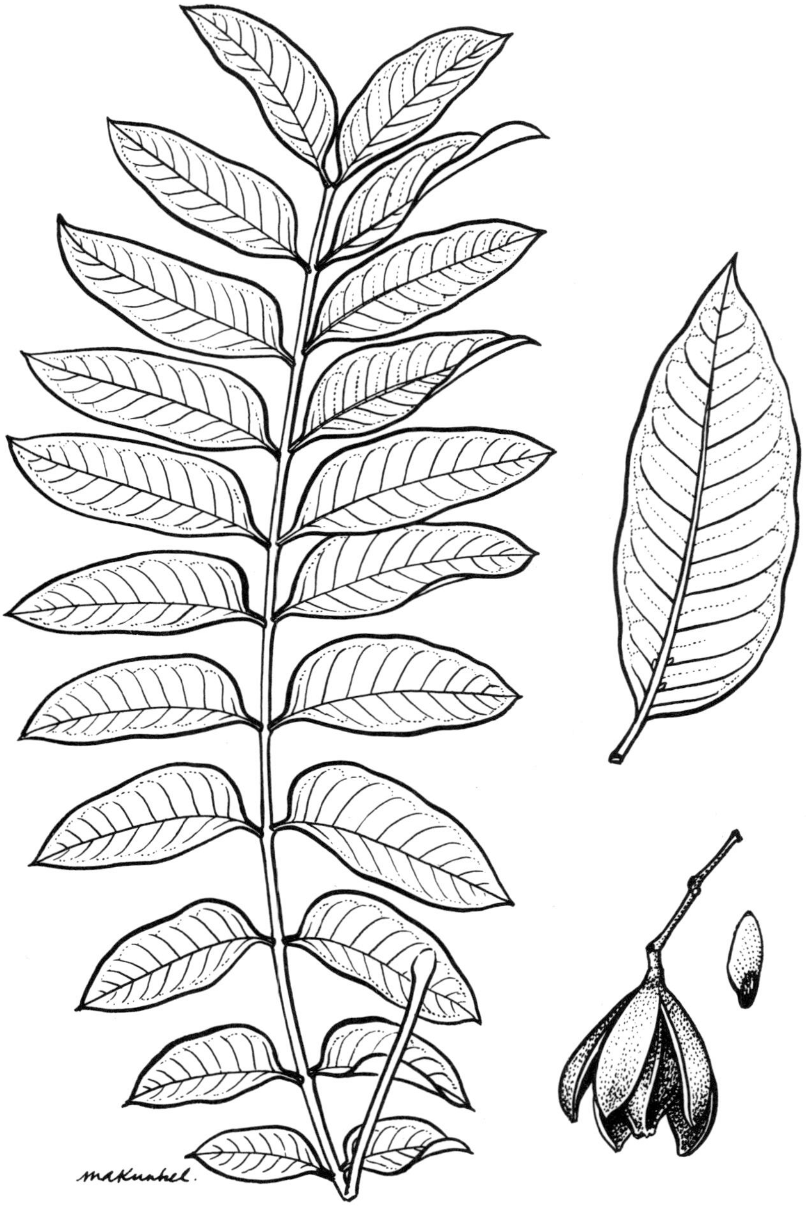

Fig. 113. *Cedrela odorata*; drawings = natural size

ACERACEAE, the Maple family

If we accept the separation of the section *Negundo* on generic level, this becomes a family of 3 genera with about 200 species of trees and shrubs mainly from northern temperate and tropical mountain zones. About 190 species belong to the genus *Acer* s.str.

Acer pseudoplatanus Linné – Sycamore

A medium-sized tree 8 to 12 (20) m tall, with a rather dense, spreading crown. Trunk more or less straight, with a strong and wide-spreading rootsystem; bark dark grey, smooth, becoming fissured, or scaled and flaky in older specimens. Branchlets chestnut coloured, with lenticels. Foliage deciduous. Leaves opposite, long-stalked; blade subcoriaceous, dark green and shiny above, palmately 5-lobed, up to 15 cm wide, with a conspicuous nerve system. Flowers greenish, in more or less terminal pendent panicles, somewhat fragrant. Fruits (samarae) winged pairs, in showy bunches.
Native in Central and Southern Europe and cultivated elsewhere except in arctic or tropical regions. Propagated from seeds and cuttings, often grown as a park or roadside tree. Wood used for furniture, carriages, gunstocks, general carpentry etc., also for fuel and charcoal. It is the Plane Maple Tree of other authors. There are several varieties, differing especially in leaf forms and their colours; 19 of these are described by Elwes & Henry who dedicate 10 pages to this particular species and cite the largest specimen known as 100 ft. tall.

Ref.: Alberti; Bean; Chittenden; Elwes & Henry; Encke; Harrison; Hart & Raymond; Lombardo 58; Menninger 64; Mitchell; Polunin; Ruiz de la Torre; Uphof; Walters.

Also in some gardens a few specimens of *Acer palmatum* Thunb. (Smooth Japanese Maple) which is a smaller tree and native in Japan, with deeply cut leaves. A very variable species, with many forms and varieties (Bean, Chittenden, Fitschen, Harrison etc.).

Fig. 114. *Acer pseudoplatanus*; drawings = 1/2

Negundo aceroides Moench – Box-Elder (Maple)

Usually cited as *Acer negundo* L. (sometimes as *Negundo fraxinifolium* Nutt.), this species is a native of North America; now widely cultivated in other temperate and subtropical zones. The genus *Negundo* differs from *Acer* s.str. by having unisexual flowers and pinnate leaves.

Tree up to 15 m in height, with a wide spreading crown. Trunk stout, bent or somewhat twisted in older specimens; bark greyish-brown, fissured or furrowed. Foliage deciduous. Leaves imparipinnate, stalked, with 3 to 5 (7) leaflets which are rather herbaceous, oblong-ovate, often unequal, coarsely dentate towards the long-pointed apex; blade bright green, up to 10 cm long and 5 cm wide. Male flowers yellowish-green, with long stamens, in slender pendent racemes; flowering before leaves appear. Fruits similar to *Acer*, in hanging racemes up to 20 cm long.

A very variable species with numerous varieties and cultivars (Bean, Harrison). A good shade and roadside tree, propagated from seeds and cuttings. According to Uphof the wood is not durable but used for paper pulp. The sap is a source of sugar. North American Indians use the charcoal for painting and tattooing their bodies. – Alternative name: Ash-leaved Maple.

Ref.: (Alberti); (Bean); (Chittenden); (Eliovson); (Elwes & Henry); (Encke); (Fitschen); (Harrison); (Lombardo 58); (Mitchell); (Ruiz de la Torre); (Uphof); (Walters).

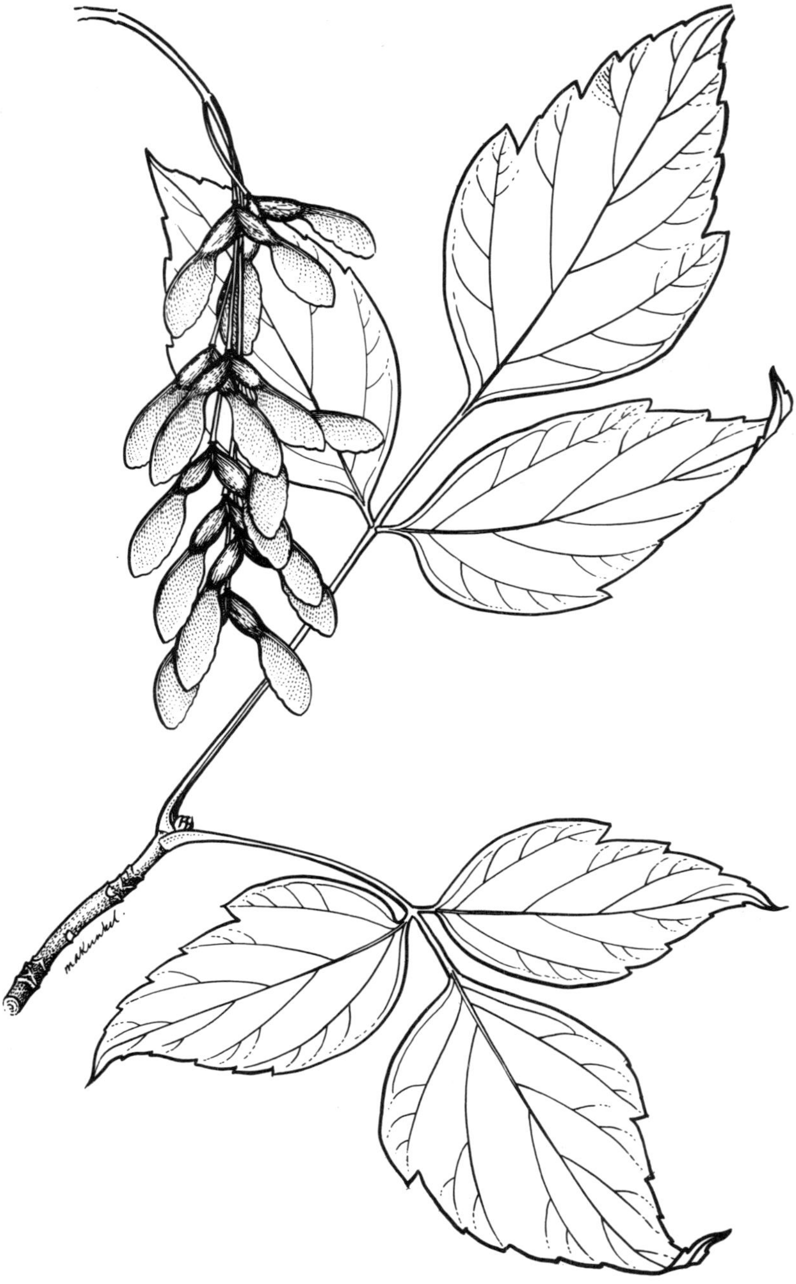

Fig. 115. *Negundo aceroides*; drawings = 3/5

SAPINDACEAE, the Soap-berry family

A large mainly tropical family of trees, shrubs and vines; about 150 genera and 2.000 species. *Blighia*, *Cardiospermum*, *Dodonaea* and others have come into cultivation as ornamentals, and the Litchi (*Litchi chinensis* Sonn.) is a fruit tree from East Asia. None of the species described below could be considered as common in the Canary Islands, and all Sapindaceae grown here are introduced.

Koelreuteria paniculata Laxm. – Golden Rain-tree

This small to medium-sized tree from China is hardy in temperate gardens. It is a gracious species with a short trunk and a wide branching open crown. Bark dark grey or brownish, fissured. Foliage deciduous. Leaves imparipinnate, 20–30 cm long, clustered on terminal branchlets; young shoots pubescent. Leaflets 9 to 15, usually sessile, ovate-lanceolate, 4 to 7 cm long, with a strongly serrate or serrately lobed margin. Flowers yellow, in showy pyramidal, terminal panicles up to 35 cm long. Fruit an inflated 3-valved capsule with pea-sized dark brown or almost black seeds.

Propagation from seeds; best grown in more temperate climates but the species also resists hot winds and drought (Bircher). A fine garden and roadside tree of which several varieties are known (Bean). According to Uphof the flowers are used in home medicine and are the source of a yellow dye. – Alternative common names: Pride of China, Pride of India and Varnish Tree.

Ref.: Barrett 56; Bean; Bircher; (Chanes); Chittenden; Eliovson; Encke; Fitschen; Graf; Harrison; Lombardo 58; Menninger 62, 64; Mitchell; Polunin & Smythies; Uphof.

Fig. 116. *Koelreuteria paniculata*; flowering branch and fruit = 2/3

Melicoccus bijugatus Jacq. – Spanish Lime

'Like the guinea pig, the Spanish Lime fails to live up completely to the two parts of its common name: it is Spanish-American, and is unrelated to the citrus fruits' (Barrett). The species is native in Central and northern South America, and is often cited as *Melicocca bijuga* L. – Alternative common names: Genip and Honeyberry.

Tree up to 20 m in height, with a rather dense crown; trunk stout, up to 80 cm in diameter. Bark grey, rather smooth. Foliage evergreen (deciduous according to Adams). Leaves alternate, pinnate, with only two pairs of leaflets; rachis winged or not so. Blade oblong-ovate, herbaceous, dark green, up to 10 cm long and 4 cm wide. Flowers small, whitish, fragrant, in spikes or panicles up to 15 cm in length. Fruits rounded, fleshy, yellowish-green, up to 3 cm in diameter, usually in clusters; pulp gelatinous and edible.

The large seeds are eaten roasted. The wood is used for cabinet work and construction; a 'decoction of bark is used in Trop. America for dysentery' (Uphof). The species is propagated from seeds and makes a good shade tree for gardens and avenues of warmer zones; it seems to be a popular fruit tree in the West Indies.

Ref.: Adams; (Barrett 56); (Bircher); (Burkill); (Chittenden); Little & Wadsworth; Long & Lakela; (Menninger 64); (Mortensen & Bullard); (Neal); (Uphof).

Fig. 117. *Melicoccus bijugatus*; leaves = 1/2

Sapindus saponaria Linné – Soap-berry Tree

An evergreen tree (deciduous in Hawaii) native in tropical America, 10 to 15 (20) m tall. Trunk slender, up to 40 cm in diameter; bark grey or somewhat brownish, warty becoming fissured or scaly. Crown dense; twigs with lenticels. Leaves alternate, paripinnate, with or without a winged rachis; leaflets 4 to 12, oblong or ovate-lanceolate, opposite or almost so, sometimes unequal, up to 16 cm long and 3 to 5 cm wide, often downy beneath. Flowers small, whitish, in open panicles up to 30 cm long. Fruit globose, yellow, poisonous, about 2 cm in diameter, usually single but with small appendages at the base which are the remains of undeveloped abortive fruits. Seed rounded, black, rather large.

The thin fleshy layer of the fruit is used as a substitute for soap. Seeds are often made into necklaces, when crushed they make a fish poison but are also said to have medicinal properties, i.e. for rheumatism. The roots also contain saponin (Burkill). The wood is used for fence posts. Otherwise planted as a shade tree. Propagation from seeds. Only for gardens at lower altitudes of warmer zones; according to a quotation cited by Menninger 'it grows luxuriantly in places where salt water overflows frequently'.

Ref.: Adams; Bircher; Burkill; Chittenden; Little & Wadsworth; Long & Lakela; Martinez; Menninger 64; Neal; Soukup; Uphof; Wiggins & Porter.

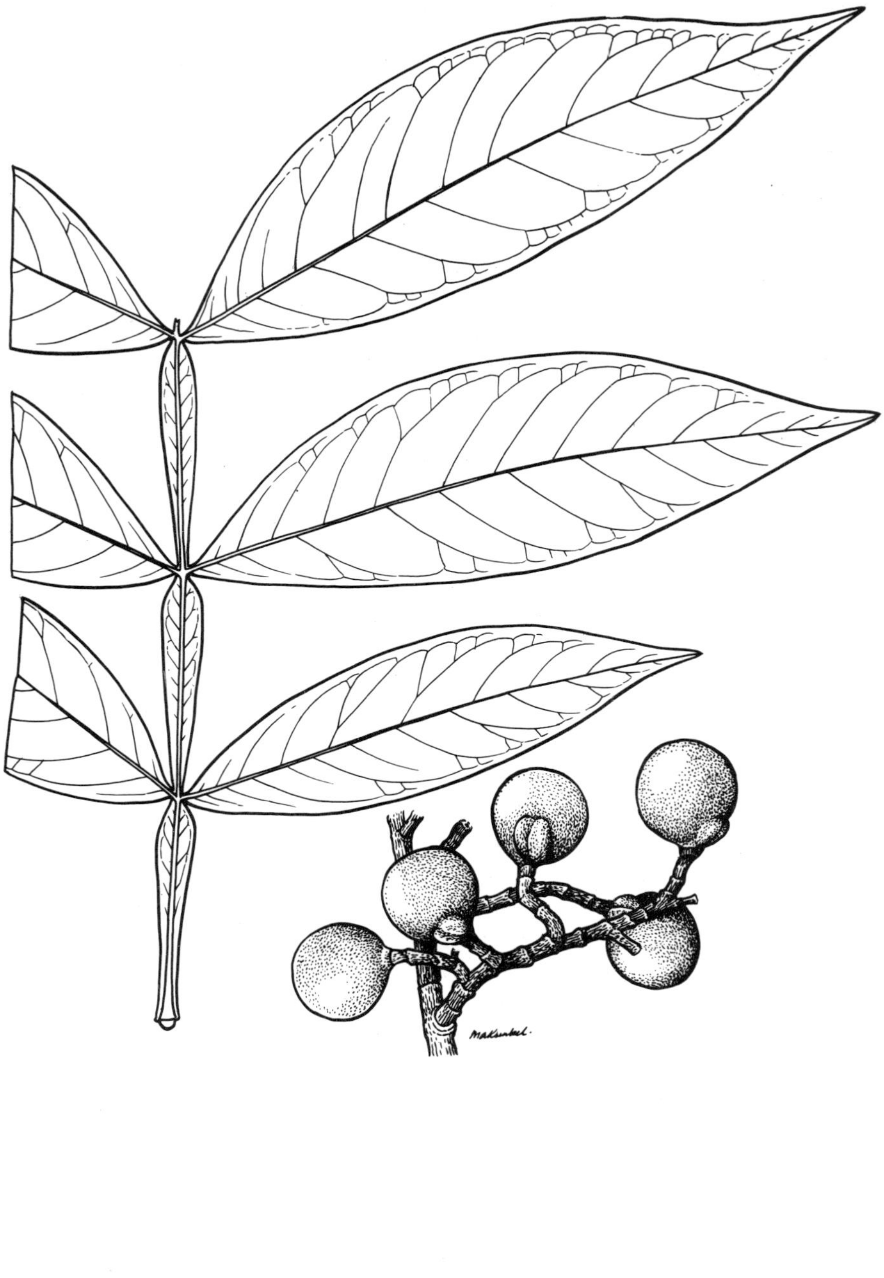

Fig. 118. *Sapindus saponaria*; leaf and fruits = 3/5

Photo 5. An Orange tree, flowering and fruiting.

ZYGOPHYLLACEAE, the Caltrop family

This family of about 25 genera with some 250 species is widely distributed along tropical and subtropical sea-shores, and most species are adapted to saline environments. Four or five species (*Zygophyllum*, *Fagonia*, *Tribulus*) are native in the Canary Islands although no endemism is present. Of interest to our account only one species:

Guaiacum officinale Linné – Lignum vitae, or Tree of Life

A small tree native in the Caribbean Islands and the Central American belt, and rare in our gardens; it is included here only because it is attractive.
From this species one of the heaviest commercial timbers known is obtained; it is said to be resistant to decay, to termites and, up to a certain point, even to the lumberman's saw. However most natural stands were cleared during the first decades following its discovery because of its supposed medicinal properties; all parts of the plant, including leaves, were shipped to European markets and alchemist's laboratories.
Tree up to 10 m tall; trunk short, often twisted or knotty. Bark greyish-brown and large-scaled; terminal branches green, with ringed nodes. Foliage evergreen. Leaves pinnate, up to 7 cm long, with minute hairy scales. Leaflets 2 (to 3) pairs, more or less obovate or almost rhomboid, dark green and glossy, herbaceous, up to 4 cm long and 2,5 cm wide. Pretty, slightly fragrant pale blue flowers in stalked, axillary clusters. The fruits are flattened brownish capsules, somewhat heart-shaped, 2 cm wide; usually 2-seeded.
A very slow growing species which has been exterminated in many original localities. Planted as an ornamental in gardens. Wood very hard, self-lubricating and resinous, used for bearings and blocks of propeller shafts of steam ships (Little & Wadsworth), in the manufacture of bowling balls, pulley blocks etc. (Neal). An extract of the wood and resin is used for medicinal purposes and in the dyeing industry. The species is recommended by Menninger for seaside planting.

Ref.: Adams; Barrett 56; Bircher; Chittenden; Irvine; Little & Wadsworth; Lombardo 58; Menninger 62, 64; Neal; Uphof.

Fig. 119. *Guaiacum officinale*; drawing = 3/4

AQUIFOLIACEAE, the Holly family

According to Willis a family of 2 genera and 400 species of trees and shrubs. in both tropical and temperate climates. *Ilex paraguariensis* St.Hil., a native of South America which produces the Paraguay Tea or Yerba Mate, is of commercial importance. – There are two native species in the Canary Islands, very variable components of the Laurisilva: *Ilex canariensis* Poir. and *I.platyphylla* Webb & Berth., sometimes seen at roadsides in the more humid higher part of the islands.

Ilex aquifolium Linné – English Holly

Evergreen tree up to 12 or 15 m tall, native in Western and Southern Europe (and some Oriental countries) but extensively planted elsewhere. Trunk erect, usually much branched or with exuberant rootshoots and suckers. Like most Hollies very variable in leaf-shape; the leaves might be almost flat and entire or undulate and strongly spiny. Leaves up to 8 or 10 cm long and 2,5 to 3 cm wide, dark green, shiny, and very tough or even brittle. Flowers whitish and fragrant, in crowded clusters. Fruits globose, shiny, scarlet, fleshy, up to 8 mm in diameter; two to four-seeded.
Ornamental as a solitary specimen, and very functional as a hedge plant. With several hybrids and cultivars grown in temperate gardens. Wood whitish, strong, heavy, easy to polish, and used for handles, sticks, wood-cuts etc.; fruiting branches much appreciated as Christmas decoration. – Bean gives for this species a height of 'up to 80 ft.', and the numerous cultivars are dealt with on four pages of his book, Only very few specimens are seen in Canary gardens. Alternative names: Common Holly, and European Holly.

Ref.: Bean; Chanes; Chittenden; Eliovson; Encke; Fitschen; Font Quer; Graf; Harrison; Lombardo 61, 68; Mitchell; Neal; Polunin & Smythies; Ruiz de la Torre; Uphof, and almost every European 'Flora'.

Fig. 120. *Ilex aquifolium*; fruiting branch = natural size

CELASTRACEAE, the Bittersweet family

A family of trees, shrubs and vines in tropical to temperate zones; about 55 genera and 850 species of which relatively few (*Catha*, *Euonymus*) are of economic importance. *Maytenus canariensis* is an endemic Canarian tree, whereas *Gymnosporia senegalensis* (Lam.) Loes., a rarity in the eastern Canary Islands, is thought to have been introduced by birds from its native home in West Africa.

Elaeodendron orientale Jacq.f. ex Jacq. – False Olive

Small tree 6 to 8 meters tall, with a slender trunk up to 25 cm in diameter. Bark dark grey, somewhat rough. Crown rather open, pyramidal, with horizontally spreading branches easily breaking off; terminal branchlets somewhat angular. Foliage evergreen. Leaves opposite-decussate, stipulate, short-stalked, oblong-lanceolate, subcoriaceous or somewhat fleshy and brittle, glabrous, shiny, 8 to 13 cm long and 2,5 to 4 cm wide; margin serrate. Flowers small, greenish, about 5 mm in diameter, in small dense clusters. Fruit globose, roughly broad olive-shaped or plum-like, greenish, fleshy, up to 2 cm in diameter; usually 1-seeded.

Propagation from seeds and cuttings. The species thrives best in moist climates at lower altitudes. It makes a good ornamental tree for smaller gardens. – Probably native in Madagascar and Mauritius. The genus is sometimes included in *Cassine*.

Ref.: Chittenden; Kunkel 69; Neal.

Fig. 121. *Elaeodendron orientale*; drawings = 3/5 except solitary flower (= natural size)

CORYNOCARPACEAE, the Karaka family

Probably related to the Anacardiaceae (in which it was frequently included), the Karaka family consists of one single genus, with 4 or 5 species. The family is native in New Zealand, New Caledonia, New Hebrides and New Guinea. In cultivation here only

Corynocarpus laevigata Forster – Karaka, or New Zealand Laurel

An evergreen tree 12 to 15 m tall native to coastal New Zealand. The pronounced but slender trunk has a smooth grey bark and may reach 50 cm in diameter. Branches somewhat knotty. Leaves alternate, stalked, dark green and glossy; obovate, thick-herbaceous to almost fleshy, up to 15 cm long and 5 cm wide. Margin entire but characterized by a visible submarginal false vein. Flowers small, whitish, in terminal panicles 15 to 20 cm long. The attractive fleshy fruit (drupe) is 2,5 to 4 cm long and 1,5 to 2 cm in diameter; fruits are orange-coloured when ripe and are very fragrant.

According to available literature the pulp is eaten raw but kernels 'have to be soaked and steamed to remove the poison' (Hamlin). Uphof reports that trunks are used by the Maoris to make canoes. – The Karaka is propagated from seeds and cuttings, and grows best in protected coastal gardens. Harrison cites several varieties of this species which is recommended by Menninger as a resistant seaside plant.

Ref.: Bircher; Chittenden; Encke; Hamlin; Harrison; Kunkel 69; Menninger 64; Neal; Uphof.

Fig. 122. *Corynocarpus laevigata*; fruiting branch = 3/5

OLEACEAE, the Olive family

A rather heterogenous family of trees and shrubs from temperate to tropical climates; over 25 genera with almost 600 species including the multitude of garden forms of *Syringa vulgaris* L., the Common Lilac, and about 300 species of *Jasminum*. *Picconia excelsa* (Ait.) DC. is a component of the Canarian Laurel forest; species of *Phillyrea* are thought to be early introductions, and *Jasminum odoratissimum* L. is a Macaronesian endemic.

Fraxinus angustifolia Vahl – Narrow-leaved Ash

A tree native in Southern Europe and North Africa, up to 15 m tall; trunk straight, bark dark grey and furrowed in older specimens. Crown spreading, younger branches drooping; buds brown. Foliage deciduous. Leaves opposite, long-stalked, imparipinnate, up to 25 cm long; leaflets 5 to 9, practically sessile, lanceolate, glabrous, dark green, distantly serrate, up to 9 cm long and 2 cm wide. Flowers inconspicuous on previous year's wood. Fruit (a samara) 1-seeded, winged, 3 to 5 cm long. The material illustrated probably belongs to var. *lentiscifolia* Henry.

Propagation from seeds and cuttings. Best grown at somewhat higher altitudes. A good roadside tree. According to Ruiz de la Torre the wood is resistant and elastic, used for cabinet-work; also for fuel and charcoal. Leaves eaten by animals; bark febrifuge. – Also cultivated but rather rare in these islands are *Fraxinus ornus* L. and *F.americana* L., present in a few private gardens.

Ref.: Bean; Bircher; Chittenden; Fitschen; Franco & Rocha Afonso; Mitchell; Ruiz de la Torre.

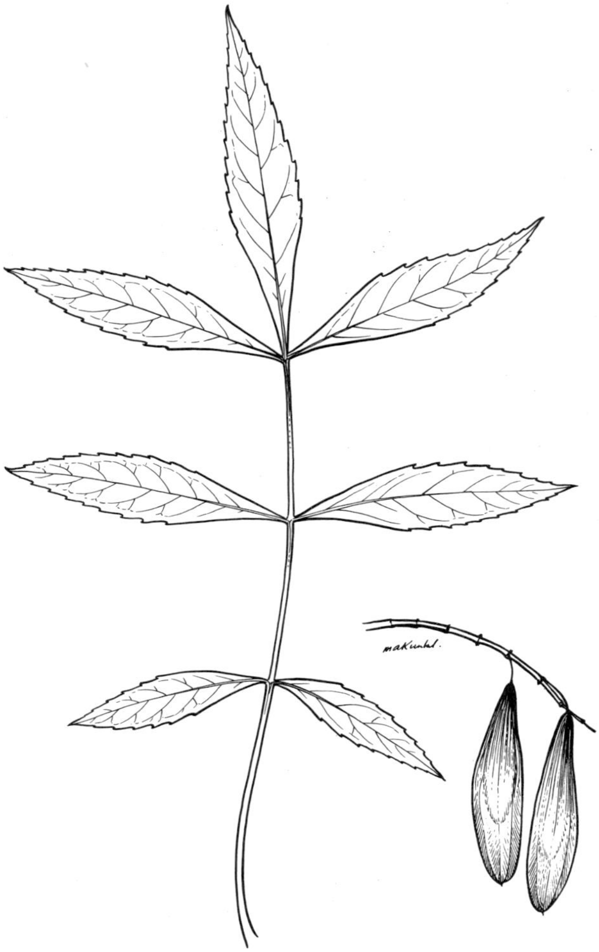

Fig. 123. *Fraxinus angustifolia* cf. var. *lentisci folia*; leaf and fruits = 1/2

Ligustrum lucidum Aiton – Glossy Privet

A small East Asiatic tree frequently seen at roadsides and in small gardens; often confused with the closely related, smaller and more rounded-leaved *L.japonicum* Thunb. – Tree 6 to 8 m high, with a pronounced but slender trunk; bark grey or dark grey, rather smooth, with lenticels. Crown dense, almost pyramidal. Evergreen. Leaves opposite-decussate, stalked, oblong-ovate, glossy dark green, thick-herbaceous, pointed at apex; blade up to 12 cm long and 6 cm wide but usually smaller. Flowers small, creamy white, fragrant, in large terminal, pyramidal panicles up to 20 cm long. Fruits fleshy, globose, blue-black, 4 to 6 mm in diameter.

An ornamental species grown from seeds and cuttings which, in these islands, thrives best in cooler zones. It is the host of the 'white-wax' insect obviously important in China (Bean, Uphof) where, according to Burkill, the fruits are thought 'to promote longevity, and are used in particular for rheumatism'. – Several varieties are known of which cv. 'Aureovariegatum' with yellow-blotched leaves is the most common.

Ref.: Bean; Bircher; Burkill; Chittenden; Encke; Graf; Harrison; Kunkel 69; Lombardo 58; Menninger 64; Mitchell; Moeller 71; Neal; Pesman; Uphof.

Fig. 124. *Ligustrum lucidum*; all drawings = 3/5

Olea europaea Linné – Olive Tree

A species almost too famous to write about. Nevertheless here is a native of the Mediterranean region which has been introduced and cultivated elsewhere. It is one of the oldest cultivated plants known and besides its economic value a fascinating species, for when walking through a plantation of thousands of Olive trees it will be seen that there are no two specimens with trunks that are alike. Personally I would not say it is a really beautiful tree but a tree with character, accustumed to suffer and which one cannot help but admire. Olive trees may reach 'biblical' ages of far over a thousand years; Hegi (who dedicated over 10 pages to this species) mentions 8 specimens still growing in the Gardens of Gethsemane thought to be over 2.000 years old, with trunks up to 2 m in diameter at breast-hight. – In Mediterranean countries there are more Olive trees than people; Spain only, as one of the most important Olive producers, is said to have over 300 million specimens (Hegi), and the same author gives data concerning approximately 300 different races or forms of Olive trees.

Usually a tree 6 to 8 m tall, with a spreading to somewhat pyramidal crown. Trunk short, bent, knotty, with ridges and hollows or otherwise deformed; bark dark grey or greyish-brown, fissured or furrowed. Evergreen. Leaves opposite-decussate, short-stalked, subcoriaceous; blade lanceolate, shiny dark green above and pale beneath, 4–7 cm long and up to 1 cm wide; pointed. Flowers small, whitish, fragrant, in axillary racemes. Fruit plum-shaped or precisely 'olive-like', their colour and size depending on the variety in question.

Propagation from seeds and cuttings; grafting is also practised. Best grown in warmer regions where regular winter rain is expected; never successful in humid tropics. Fruits are eaten preserved in vinegar or a salty liquid, and are the source of the most important of all oils; 'also used for lubrication, soap, formerly a base for perfume, for lights, and anointment' (Neal). For medicinal purposes considered 'as laxative, emolient, and demulcent' (Uphof); also an aperitif and tonic. The tree produces a hard and noble wood used for tools and cabinet work, and the leaves (or leafy branches) since pre-Christian times have been considered as a symbol of peace. – In the Canary Islands grows a form (probably native) known as ssp. *cerasiformis* (Webb & Berth.) Kunkel & Sunding.

Ref.: Barrett 56; Beyron; Bircher; Burkill; Chanes; Chittenden; Cobley; Encke; Esdorn & Pirson; Font Quer; Franco & Rocha Afonso; Graf; Grant & al.; Hegi V/3; Huxley; Lombardo 58; Menninger 64; Mortensen & Bullard; Neal; Polunin & Huxley; Rauh; Ruiz de la Torre; Schaeffer; Schütt; Uphof.

Fig. 125. *Olea europaea*; flowering branch and fruits = 2/3

PROTEACEAE, the Protea family

A remarkable family of trees and shrubs mainly from the drier climates of the southern hemisphere: about 60 genera with over 1.000 species, especially in South Africa (*Protea*, *Leucadendron*) and Australia (*Hakea*, *Grevillea*, *Banksia*, etc.). According to 'Flora Europaea' (1: 69) two species of *Hakea* have been planted in the Iberian Peninsula where they are now well naturalized. In the island of Madeira many Proteaceae are grown for the export of flowers to European markets.

Grevillea robusta A.Cunn. – Silver Oak, or Silk Oak

An attractive ornamental and shade tree from Australia, common in Canary gardens and widely distributed in other subtropical countries. However in Puerto Rico no longer recommended because 'the trees are heavily attacked by scale insects, and the silky foliage becomes dirty' (Little & al.).
An evergreen or semideciduous tree up to 20 m in height, in its home country up to 50 meters. Trunk straight, in our cultivated specimens usually up to 50 cm in diameter. Bark relatively thin and greyish, fissured and becoming furrowed when older. Crown more or less pyramidal and quite open, often deformed by prevailing winds. Leaves 'fern-like', pinnate or almost bipinnate, up to 25 cm long, subcoriaceous; they are dark glossy green above and silky white beneath; rachis slightly tomentose. Flowers of rich golden colour, unilaterally placed forming comb-like racemes. Capsules pod-like, flattened, dark brown to almost black, with 1 or 2 winged brown seeds.
Propagation from seeds; saplings fast growing. Wood light, brown-coloured, used for furniture and cabinet work but is attacked by termites. The species is a favoured pot-plant in cooler climates.

Ref.: Adams; Barrett; Bircher; Burkill; Chanes; Chittenden; Corner; Eliovson; Encke; Graf; Harrison; Kunkel 69; Little & al.; Lombardo 58; Long & Lakela; Menninger 62, 64; Moeller 71; Neal; Perry; Uphof.

A second species of this genus and obviously rare in cultivation is *Grevillea nematophylla* F. v. Muell.

Fig. 126. *Grevillea robusta*; drawing = 3/5

Grevillea nematophylla F.v.Muell. – Thread-leaved Grevillea

The material was kindly named by the Staff of The Royal Botanic Gardens, Kew, and I am unable to cite any references. The species is included in this book for its remarkable flowers, its strange growth form, and its resistance to drought, making a showy and easy growing ornamental. It is a native of Australia which was introduced some years ago to these islands, and might be a small tree in its home-country. It has very stiff, needle-like twigs of glaucous or greyish-green colour and seems to be always leafless. The small, pale or greenish-yellow flowers appear in more or less terminal, upright spikes which are much visited by bees. Fruit-capsules dark brown, one to three-seeded. The species makes an excellent garden tree and could well be recommended for planting on semi-arid slopes.

Fig. 127. *Grevillea nematophylla*; flowering branch and fruit = 3/5, flower details = × 3,5

Macadamia integrifolia Maiden & Betcke – Queensland Nut

Usually cited as *M.ternifolia* F.v.Muell., this well-known species of our gardens must be recognized as *M.integrifolia* as pointed out by C. C. Townsend (in litt.), a decision taken after a critical revision of cultivated material.
Small-sized evergreen tree up to 10 m tall, with a slender trunk and a small, dense crown. Bark grey. Leaves in whorls; short-stalked, narrow-oblong or lanceolate, coriaceous, somewhat undulate, greyish green and up to 15 (20) cm long; apex strong pointed. Flowers very small, white, fragrant, in narrow racemes 12 to 15 cm long. Fruits (nuts) usually one-seeded, rounded, up to 2,5 cm in diameter, with a thick, extremely hard shell which is brownish and shiny. Seeds edible, either raw or roasted and salted. According to Neal the kernels 'contain about 70 percent fat, are a good source of vitamin B_1, calcium, phosphorus, and iron; seven of them yield as many calories as one and a half slices of bread'.
The Queensland Nut is planted for its fruits, for its ornamental value especially in smaller gardens, and for reafforestation purposes. Apparently propagated from seeds and cuttings; however special attention is drawn to an extensive paper by W. B. Storey, dealing with several aspects of cultivation of this species.

Ref.: (Barrett); (Bircher); (Chittenden); (Degener); (Encke); (Esdorn & Pirson); (Graf); (Harrison); (Kennard & Winters); (Little & al.); (Lombardo 58); (Moeller 68); Mortensen & Bullard; (Neal); (Perry); (Purseglove); (Storey); (Synge); (Uphof).

A third and more recently introduced genus is represented by

Stenocarpus sinuatus Endl. – Fire-wheel tree

A tree from Southern Australia which, according to Chittenden may reach 30 m in height. Leaves deeply cut or pinnately lobed, up to 40 cm long. Flowers bright red, forming very showy dense umbels. – Not recommended for cultivation in extra-arid zones.

Ref.: Bircher; Chittenden; Encke; Graf; Harrison; Menninger 62.

Fig. 128. *Macadamia integrifolia*; all drawings = 1/2

APOCYNACEAE, the Dogbane family

A highly diverse family of trees, shrubs, herbs and vines mainly of tropical climates; there are even a few succulents in this family. Over 175 genera, with some 1.500 species, mostly with a whitish and more or less poisonous latex. Many ornamental species used in gardens; others (*Catharanthus*, *Rauvolfia*, *Vinca*, *Strophanthus* etc.) are important to medicinal industries, and forest trees such as *Alstonia*, *Funtumia* and *Landolphia* yield commercial rubber.

Plumeria rubra Linné – Frangipani

The Frangipani is frequently wrongly named, and names such as *Plumeria alba*, *P.acutifolia* and *P.acuminata* usually apply to this one particular species (Grant & al.). Even 'P.tricolor' of some garden books may again refer to the same taxon under discussion. *Plumeria rubra* seems to be a native of Central America.

A small tree with a short, slender trunk and wide spreading branches; up to 6 m tall. Bark of stems greyish-green, thin, showing scars of earlier leaf insertion. Young branches are green and shiny, and are almost succulent. Deciduous when kept under somewhat natural conditions; when regularly watered more or less evergreen. Leaves in terminal clusters, stalked, dark green, herbaceous to almost fleshy, with a pronounced ivory-coloured venation; oblong-oblanceolate and 30–40 cm long, showing a definite secondary 'margin'. Flowers whitish, cream-coloured or rose, up to 5 cm in diameter, in large, showy, fragrant terminal cymes. Fruits (follicles) in pairs on a common stalk, more or less cylindrical and up to 20 cm long.

The Frangipani is a famous flower-tree of Hawaii; the waxy flowers are made into collars to welcome visitors or given at the moment of the inevitable 'aloa'. The tree's milky juice has medicinal properties but is poisonous in larger doses. According to Burkill a decoction of the bark is given in Indonesia for veneral diseases, and the milk is purgative and used as a counter irritant for toothache or when dropped into sores. The plant is used to cure intestinal disorders in horses, and became sacred in India and Sri Lanka where it is commonly found planted around temples. The tree is easily propagated from cuttings and thrives best at lower altitudes.

Ref.: Barrett 56; (Beyron); Bircher; (Burkill); Chittenden; (Corner); Eliovson; Encke; Graf; Grant & al.; Harrison; (Irvine); Little & Wadsworth; Lombardo 61; (Martinez); Marzocca 52; (Menninger 62, 64); (Moeller 68); Neal; Nowicke; Perry; Pesman; Standley & Williams 69; (Uphof).

Fig. 129. *Plumeria rubra*; drawings = 1/2

Thevetia peruviana (Pers.) K.Schum. – Yellow Oleander

This shrub or small tree from dry tropical America is often cited as *Thevetia neriifolia*, a synonym given for the resemblance of its Oleander-like leaves. It is widely cultivated in tropical and subtropical zones, grown as a solitary specimen or in hedges which tolerate frequent pruning; it will also withstand a prolonged dry season. The seeds are poisonous but Neal records that, in the West Indies, these are carried as talismans and are known as 'lucky seeds'. Evergreen shrub or small tree up to 6 m in height, with a slender trunk and an open, wide-spreading crown. Bark thin and dark grey; branches shiny green, with leaf scars. Leaves linear-lanceolate, up to 15 cm long and 1 cm wide, herbaceous, glossy-green above and pale green beneath; strongly 1-veined, with a small and fine lateral venation. Large yellow flowers, solitary or grouped, long-stalked, tube-shaped, with petals up to 6 cm long. Fruits long-stalked, hard when ripe, forming a somewhat triangular swollen bag wider than long, which contains 1 or 2 large seeds.

The species is propagated from seeds and cuttings. Both Uphof and Burkill record that the bark contains a powerful antiperiodic and febrifuge, and it is also reported that the wood is a fish poison. Neal calls this species the 'Be-still tree', and Irvine the 'Exile Oil tree'; according to the latter author seeds were used by American Indians to produce a poison, and some varieties still serve the same purpose in India. For further information see Burkill. *Thevetia* is recommended for sea-side planting by Menninger.

Ref.: Bircher; Burkill; Chittenden; Corner; Degener; El Hadidi & Boulos; (Eliovson); Encke; Graf; Grant & al.; Irvine; Kunkel 69; Little & al.; Lombardo 61; Long & Lakela; Martinez; Marzocca 52; Menninger 64; Neal; Nowicke; Pesman; Soukup; Standley & Williams 69; (Uphof).

Fig. 130. *Thevetia peruviana*; drawings = 3/5

BIGNONIACEAE, the Trumpet-flower family

A large family of over 100 genera and some 650 species of trees, shrubs, vines and a few herbs, native in tropical and subtropical regions. The family has provided some fine ornamentals, including climbers such as *Pyrostegia venusta* (from S.America), *Arrabidaea magnifica* (S.America), *Pithecoctenium cynanchoides* (S.America), *Phaedranthus buccinatorius* (C.America), *Pandorea jasminoides* (Australia), *Podranea ricasoliana* (S.Africa) and others, all of which are cultivated in these islands.

Catalpa bignonioides Walt. – Catalpa, or Indian Bean

A small to medium-sized deciduous tree native in eastern North America and much cultivated in temperate gardens. The Catalpa reaches 10 to 12 (15) m in height; trunk relatively slender, bark greyish-brown, somewhat scaly. Crown much branched and spreading. Leaves opposite or in whorls, long-stalked; blade broad ovate or almost heart-shaped, up to 25 cm long and about 20 cm wide, herbaceous to subcoriaceous; margin entire or slightly lobed. Leaves glabrous above and somewhat downy beneath. Flowers more or less bell-shaped with a lobed or fringed margin, white with yellow lines and purple spots, in large and showy pyramidal panicles. Fruit capsules narrow, almost cylindrical, black, up to 35 cm long.

Propagation from seeds and cuttings. Wood durable, used as fence posts and railway sleepers. The species makes a good roadside tree. 'No large garden or park ought to be without one or more specimens' (Bean). Several cultivars and hybrids are also known.

Ref.: Bean; Chittenden; Eliovson; Encke; Fabris; Fitschen; Graf; Harrison; Lombardo 58; Mitchell; Neal; Perry; Uphof.

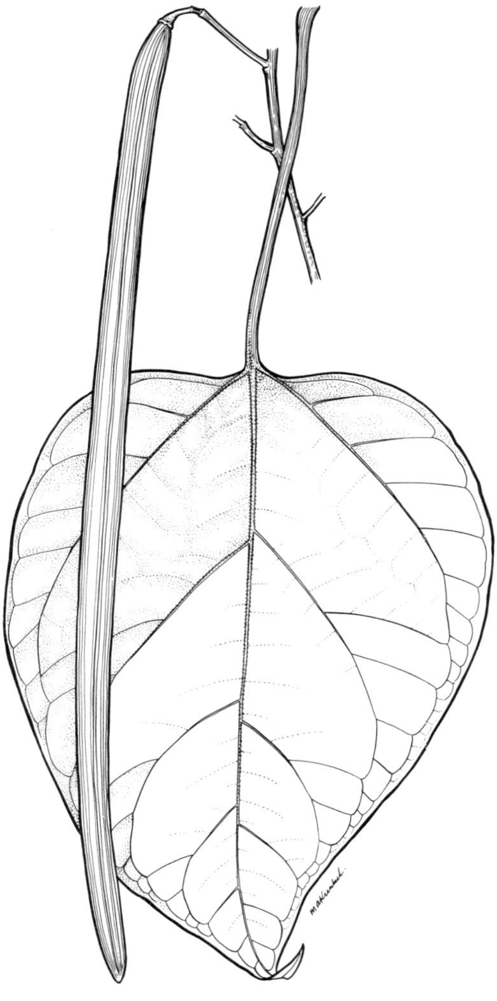

Fig. 131. *Catalpa bignonioides*; leaf and fruit = 2/5

Crescentia cujete Linné – Calabash Tree

Perhaps one would not say that this is a beautiful tree but it is certainly a curious one. The Calabash Tree is native in tropical America. There are some specimens in Canary gardens at lower elevations.

A tree up to 10 m tall, usually with a short and crooked stem up to 50 cm in diameter; bark dark grey, deeply fissured or somewhat scaly. Branches upright or spreading, extremely knotty, and easily breaking off. Evergreen or semi-deciduous. Leaves simple, herbaceous, oblong or oblanceolate, clustered; up to 15 cm long and 5 to 6 cm wide, with a panduriform venation. Flowers irregularly lobed, single or in groups, whitish-green with purplish spots and veins, appearing on the trunk (cauliflorous) or on branches and having a musty smell. The fruit is a hard-shelled 'calabash', oval-shaped, up to 30 cm long and 20 cm in diameter, which contains a blackish, jelly-like pulp embedding the numerous flattened seeds.

Propagation from seeds and cuttings. The pulp is said to be edible but one must be very hungry to face it. Uphof says that it is used in home medicine as a laxative, astringent and expectorant. Young fruits are prepared as pickles, and seeds are eaten cooked. Martinez mentions that the pulp is used in Mexico for coughs, and Burkill reports that a decoction of the bark is effective to clean wounds, and that the pounded leaves are applied for headaches. African tribes prepare the seeds as an antidote for snake-bites (Irvine). The tree is recommended by Menninger for seaside gardens. The most common use however refers to the dried, hard-shelled fruits which can be cleaned to serve as recipients and drinking vessels, or are decorated on the outside and sold as souvenirs.

Ref.: Adams; Barrett 56; Bircher; Burkill; Chittenden; Gentry; Irvine; Kunkel 69; Little & Wadsworth; Long & Lakela; Martinez; Menninger 64; Neal; Perry; Standley & Williams 74; Uphof.

Fig. 132. *Crescentia cujete*; all drawings = 3/5

Photo 6. *Crescentia cujete*, view of branches.

Jacaranda mimosifolia D.Don – Jacaranda

Probably the best known Tree Bignonia in cultivation. And a much confused one also, at least when referring to its scientific name: it is the *Jacaranda ovalifolia* R.Br. of some authors, and *J.acutifolia* of others (but not of H.B.K.). 'The naming of this species represents a fantastic coincidence in which the same plant was described in two different publications under two different names on the same day (June 1, 1822). Don was the first to unite *J.ovalifolia* with his *J.mimosifolia* so his choice of name must be accepted' (Gentry, p. 864). The Jacaranda is native in drier regions of tropical South America (S.Brazil, NE.Argentina, N.Uruguay), and is now cultivated in most countries with a suitable climate, from Hawaii to South Africa and New Zealand. – Common in Canary gardens.

Deciduous tree up to 15 m in height. Trunk short, often curved, 40 to 60 cm in diameter; bark grey, somewhat rough and deeply fissured in older specimens. Branches upright or spreading forming a wide crown. Leaves opposite, bipinnate, 30 to 40 cm long; in young specimens up to 60 cm. Pinnae opposite, 12 to over 20 pairs, each with 9 to over 30 small, ovate and sharply pointed leaflets. Flower bell-shaped, bluish-violet, up to 5 cm long, in showy open panicles. The fruit is a woody, brown, flattened, suborbicular capsule which contains many flat and winged seeds.

The Jacaranda is grown from seeds or propagated from woody cuttings. Even larger specimens, heavily pruned, stand transplanting. It makes an excellent solitary and a good roadside tree from sea-level to almost 1.000 meters above. Unfortunately it is considered to be 'dirty' because it continually sheds either flowers or leaves and then the leaf-stalks. The wood is compact and durable, used for carpentry. Fruit-capsules or bunches of these are often employed for decorative purposes. The species is sometimes also called 'Rosewood tree', and a white-flowering variety is cited by Chittenden. Standley & Williams, in their partial treatment of the 'Flora of Guatemala', are very enthusiastic about this species: 'Whoever loves jacaranda trees will be delighted with Guatemala, for in few regions is one very far from at least a small number of them. One jacaranda tree is a lovely sight, never to be forgotten, but hundreds of thousands rather pall upon one ...'.

Ref.: Adams; (Barrett 56); Beyron; (Bircher); Chanes; (Chittenden); (Degener); (El Hadidi & Boulos); (Eliovson); (Encke); Fabris; Gentry; (Graf); Harrison; Irvine; (Kunkel 69); Little & Wadsworth; (Lombardo 58); (Long & Lakela); Menninger 62; Moeller 68; (Neal); Perry; (Pesman); Schaeffer; (Soukup); Standley & Williams 74; Synge; Uphof.

Fig. 133. *Jacaranda mimosifolia*; drawings = 3/5

Kigelia africana (Lam.) Benth. – Sausage Tree

The reason for the popularity of this African tree probably was not for its flowers but for the curious, sausage-like fruits. Although some authors describe several species of this genus, it is now believed (according to Kew, in litt.) that all the described taxa should be united under one single name: *Kigelia africana*, with *K.pinnata* (Jacq.) DC. as its most common synonym.

A medium-sized tree up to 15 m tall, with a spreading crown. Trunk up to 60 cm in diameter, bark grey or pale brown. Foliage semi-deciduous. Leaves alternate, imparipinnate, up to 25 (30) cm long. Leaflets usually 7 to 11 but sometimes without the terminal one; more or less opposite, leathery, ovate, oblong or obovate and up to 12 cm long; apex pointed, rounded or slightly emarginate. Flowers irregular, cup-shaped and bent, dark purple with greenish lines, in hanging panicles; their smell is unpleasant. Fruit cylindrical or slightly adpressed, pale grey, up to 80 cm long and 10 cm in diameter, hanging on long string-like stalks and containing a hardish pulp and many small seeds.

The fruit, in spite of its appearance, is not edible but the pulp is used medicinally. Uphof says that it has purgative properties, whereas Palmer & Pitman state that seeds are eaten roasted, and the baked fruit is used for fermenting beer; the fruit is also employed as a dressing for ulcers, as an enema to cure children's stomach ailments, and to treat veneral diseases and rheumatism. Wild animals consume flowers and leaves. Trunks are used for making canoes.

Ref.: Aubreville; (Barrett 56); Bircher; (Burkill); (Eliovson); (Gentry); (Graf); Irvine; Moeller 71; (Neal); Keay & al.; Palmer & Pitman; Perry; Uphof; Van der Spuy.

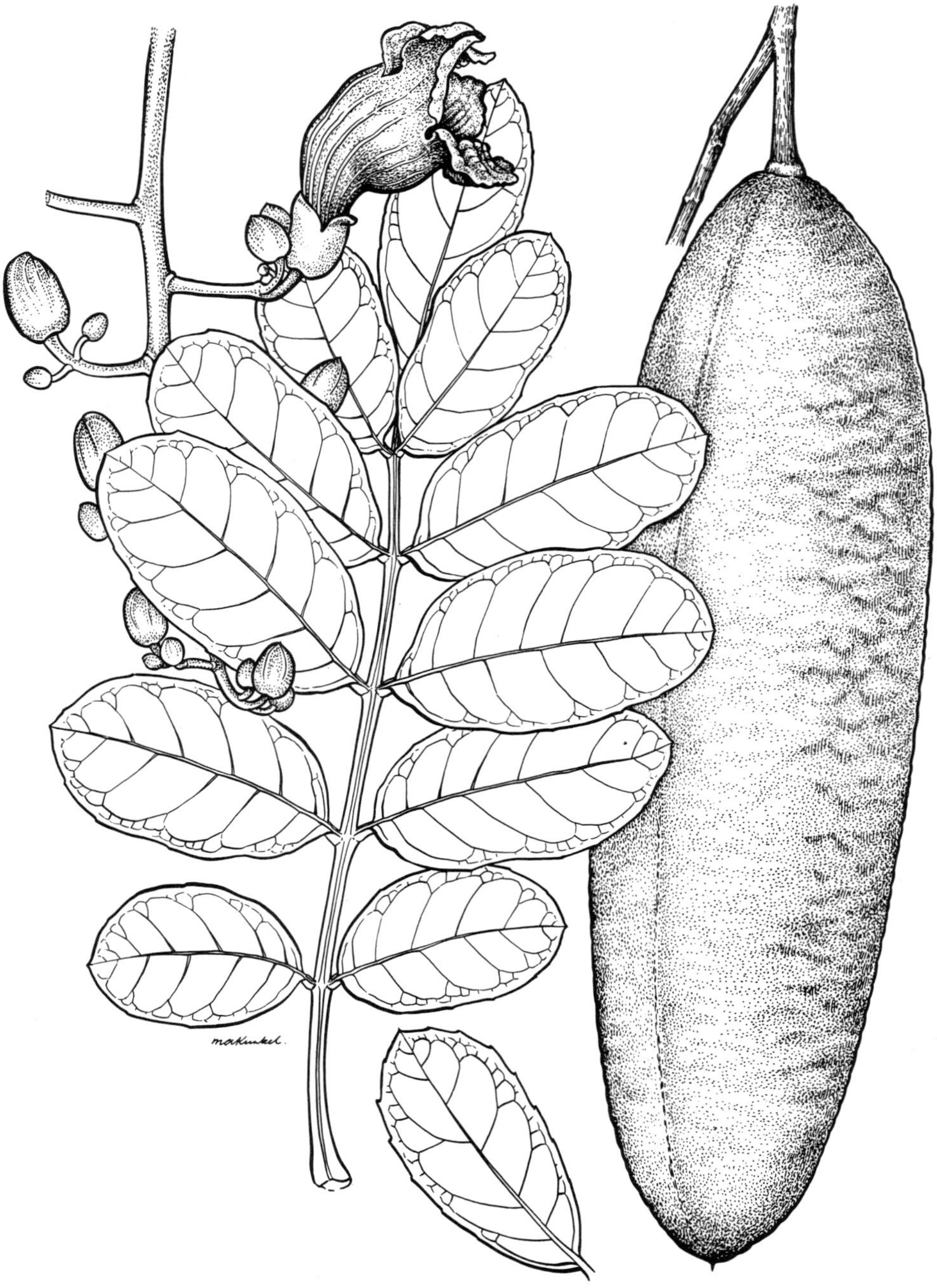

Fig. 134. *Kigelia africana*; all drawings = 3/5

Spathodea campanulata P.Beauv. – African Tulip Tree

A really handsome evergreen tree from tropical Africa and now planted extensively in other tropical and subtropical countries as an ornamental and shade tree.
Tree 15 to 25 m in height, with a well-formed trunk 40 to 60 cm in diameter. Bark greyish-brown, fissured or (in older trees) scaly. Branches upright forming a dense, more or less rounded dark crown; twigs with lenticels. Leaves opposite, imparipinnate, 20 to 40 (50) cm long. Leaflets 7 to 17, oblong or obovate, acuminate, dark green, herbaceous and up to 10 cm long; slightly puberulous on the rachis and secondary veins. Flowers roughly cup-shaped, red-orange with pale and finely crenate borders, usually up to 5 cm across; upright, in very showy, crown-like, dense terminal racemes normally overtopping the leaf-level. Fruit-capsule oblong-lanceolate, flattened, up to 20 cm long, splitting open in two boat-shaped valves to release numerous thin-winged seeds.
Cultivated best at lower altitudes and propagated from seeds and cuttings. According to Neal, 'as unexpanded flowers contain water, it is also called the 'fountain tree''. Irvine suggests that seeds are edible. Furthermore, according to the same author, an infusion of the bark is used as an enema for backaches, and pulp of bark to treat skin diseases and dysentery. A leaf decoction may be used as a poison antidote, for veneral diseases and kidney trouble. – Freshly cut flowers keep well in water, and I suggest that while observing them, in an intense and concentrated manner, they may even cure mental perplexity!
According to Menninger (1962) there exists a form with pure yellow flowers. Eliovson also records some different names for this tree: Fiery torch, and Nondi flame. – 'The tree is one of the finest ornamental trees introduced in Central America' (Standley & Williams).

Ref.: Adams; Aubreville; Barrett 56; Beyron; Bircher; Chittenden; Corner; Degener; El Hadidi & Boulos; Eliovson; Gentry; Graf; Harrison; Irvine; Keay & al.; Kunkel 69; Little & Wadsworth; Menninger 62, 64; Moeller 68, 71; Neal; Perry; Pesman; Schaeffer; Standley & Williams 74; Van der Spuy.

Fig. 135. *Spathodea campanulata*; flowering branch = 3/5, fruit = 1/2

Tabebuia rosea (Bertol.) DC. – Trumpet Tree

Although this genus may consist of almost a hundred species, natives of more tropical parts of America from Mexico to Argentina, only few are really widely cultivated in our gardens. One of the more common species is *Tabebuia rosea*, often cited under its synonym *T.pentaphylla* (L.) Hemsl. which again is kept separate from the former, by Standley & Williams.

In its native environment this species grows into a sizeable tree up to 20 meters in height, with a trunk up to 1 m in diameter; it is usually a much smaller tree in gardens. Bark dark grey and deeply fissured; branches with lenticels. Foliage deciduous or semi-persistant. Leaves long-stalked; leaflets usually five, but showing no liking for rules, there may be any number from one to five; oblong-ovate, almost coriaceous and dark green. Flowers in terminal clusters, pinkish-white with darker lines, 5 to 8 cm across when fully open, corolla lobed, finally hairy with a frilled margin. Fruit capsules linear-cylindrical, dark grey, 15 to 20 cm long; numerous flattened winged seeds.

Propagation from seeds and cuttings. The species thrives best at lower altitudes. According to Gentry a good timber tree. A second species cultivated in these islands, with paler flowers also with an unsettled number of leaflets is probably *T.pallida* (Lindl.) Miers.

Ref.: (Barrett 56); (Bircher); Chittenden; Gentry; Graf; Irvine; (Kunkel 69); (Little & Wadsworth); (Long & Lakela); (Martinez); Menninger 60; (Neal); Perry; Standley & Williams 74(?); (Uphof).

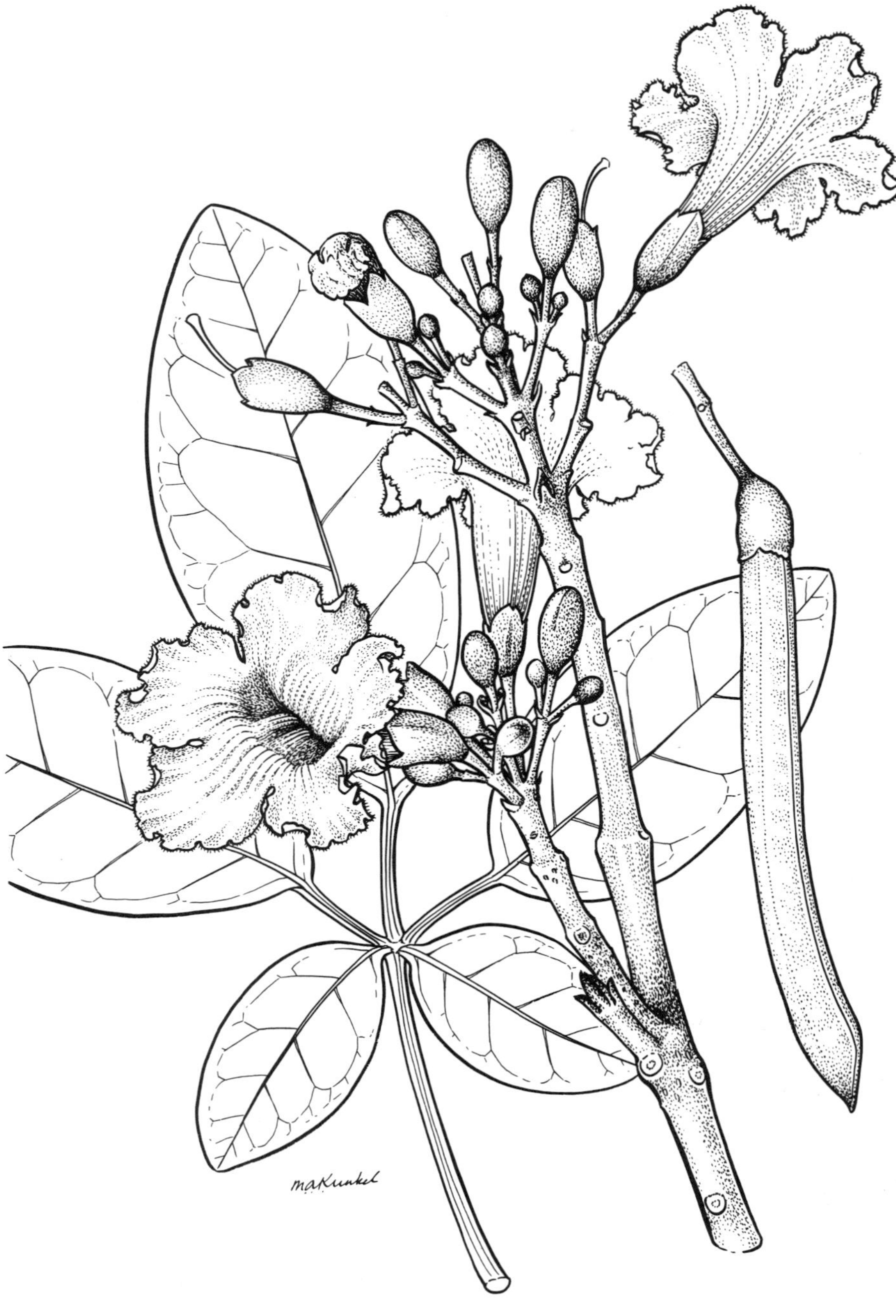

Fig. 136. *Tabebuia rosea*; drawings = 3/5

MYOPORACEAE, the Myoporum family

Of the 4 or 5 genera and 90 species recognized (Adams: 180 spp.!), most are natives of Australia, New Zealand, and other South Pacific Islands. Most species are shrubs or small trees, a number of these (mainly belonging to the genus *Myoporum*), being used for seaside planting.

Bontia daphnoides L. – Mangle Bobo

A small tree 6 to 8 meters tall apparently native in the Caribbean region and one of the very few New World members of this family. Trunk slender, erect, 30 to 40 cm in diameter; bark pale brown, rough or furrowed. Younger branches with leaf-scars. The species seems to be evergreen. Leaves lanceolate, up to 10 (12) cm long, pointed at the apex; blade with fine glands showing against the light. Flowers solitary, axillary and stalked, somewhat tube-shaped with a down-curved lower lip which is purplish-hairy on the showing side; otherwise the corolla is more or less yellowish or cream-coloured, up to 2 cm long. Fruits (drupes) ovoid-globose, hard-fleshy, greenish, up to 1 cm in diameter; each fruit contains one single thick-walled seed.

The species is recommended for seaside gardens but grows also at higher altitudes. Besides its ornamental value no other uses are known. Plants are usually propagated from seeds; occasional cuttings also strike.

Ref.: Adams; Bircher; Little & al.

Fig. 137. *Bontia daphnoides*; branch and leaf = 3/5, flower and fruit detail = × 1,2

VERBENACEAE, the Verbena family

About 75 genera with over 3.000 species, most abundant in tropical regions. The family consists of herbs, shrubs, trees and vines of which species of *Clerodendron*, *Petrea*, *Phyla* and *Verbena* are common garden ornamentals. Also *Lantana camara* L., an uni- or multicoloured flowering shrub which, unfortunately, becomes too frequently a 'weed'. *Tectona grandis* L.f. from Southeast Asia is the source of the famous Teak wood. *Vitex agnus-castus*, the Chaste tree, and *Lippia triphylla*, the Lemon Verbena, will be dealt with in a second volume of this series. Quite common in Canary gardens is

Citharexylum spinosum Linné – Fiddlewood

A West Indian tree often recorded as *C.quadrangulare* Jacq. which seems a more appropriate name as our specimens are not spinous. However the Linnean name has priority and, therefore, must be accepted. The Fiddlewood or 'Susanna' is mainly grown as an ornamental and is most successful near to the coast. Regular pruning is recommended in order to obtain richly flowering specimens and to avoid the frequent attacks by scale insects and white-woolly lice which occur in these islands.

Tree up to 15 m in height but usually smaller, with a short slender trunk up to 25 cm in diameter; bark dark grey and furrowed. Branches upright forming a more or less pyramidal crown; young branches 4-angled, easily breaking off. After a drastic pruning young branches may grow 4 to 5 meters during one growing season. Foliage deciduous and only semi-persistant when regularly watered. Leaves opposite-decussate, petiolate, oblong-ovate, with a blunt or pointed apex; herbaceous or subcoriaceous, intense green and up to 18 cm long; they usually turn reddish after the flowering season. Flowers small, white and very fragrant, in narrow axillary or terminal racemes up to 30 cm long. Fruits globose, orange-coloured or purplish, somewhat fleshy, 6 to 8 mm in diameter.

Uses unknown except for its ornamental value. However Uphof reports that the wood serves for general building purposes, and 'for guitars used by natives'. Corner thinks that the name 'fiddle-wood' might be a corruption of the French 'Bois fidele' meaning the reliability of the timber.

Ref.: Bircher; (Chittenden); (Corner); (Kunkel 69); Little & al.; Neal; (Uphof).

Fig. 138. *Citharexylum spinosum*; flowering branch and fruits = 3/5, flower details and solitary fruits = natural size

Duranta repens Linné – Golden Dewdrop

A small and variable tree or shrub up to 7 m tall native in tropical and subtropical America, from Florida and California down to Argentina and Uruguay; cultivated elsewhere for its showy flowers and fruits.

Trunk slender, up to 20 (25) cm in diameter; bark dark grey and fissured. Branches upright or arching, unarmed or somewhat spiny, often semi-scandent. Twigs more or less 4-angled. Usually evergreen. Leaves opposite, herbaceous, glabrous, ovate-elliptic or obovate, up to 5 cm long and 3 cm wide; margin entire or coarsely serrate towards the obtuse or acuminate apex. Flowers bluish or lilac (some varieties with white or purple flowers), fragrant, up to 1,5 cm across, in axillary or terminal racemes or panicles up to 20 cm long. Fruits globose and fleshy, orange-yellow, about 1 cm in diameter.

The species is usually propagated from cuttings and thrives best at lower altitudes. It makes a good hedge or, when shaped, an admirable solitary plant. According to Martinez the fruits are febrifuge, and to the flowers are attributed similar properties. Little & al., on the other hand, report that the fruits are poisonous to humans, whereas leaves serve for home remedies.

Ref.: Adams; Bircher; (Chittenden); Corner; Eliovson; Encke; Graf; (Harrison); Irvine; Kunkel 69; Little & al.; Lombardo 64; Long & Lakela; Martinez; Moldenke; Neal; Perry; Pesman.

Fig. 139. *Duranta repens*; drawings = 2/3

Common names – I

(English or American names, as cited in the text)

Acacia – *Robinia pseudacacia*
African Tulip Tree – *Spathodea campanulata*
Albizia – *Albizia lophantha*
Alligator Pear – *Persea americana*
Allthorn Acacia – *Acacia karroo*
Almond Tree – *Amygdalus communis*
American Lime – *Tilia americana*
American White Elm – *Ulmus americana*
Anón – *Annona squamata*
Apple – *Malus pumila*
Apricot – *Armeniaca vulgaris*
Aroeira – *Schinus terebinthifolius*
Ash-leaved Maple – *Negundo aceroides*
Aspen – *Populus tremula*
Atemoya – *Annona cherimolia × squamosa*
Australian Banyan – *Ficus macrophylla*
Australian Blackwood – *Acacia melanoxylon*
Australian Kino – *Eucalyptus camaldulensis*
Australian Pine – *Casuarina equisetifolia*
Australian River Oak – *Casuarina cunninghamiana*
Australian Weeping Wattle – *Acacia saligna*
Avocado Tree – *Persea americana*
Bailey Wattle – *Acacia baileyana*
Balsa – *Ochroma pyramidale*
Banjo Fig – *Ficus lyrata*
Banyan Fig – *Ficus benghalensis*
Baobab – *Adansonia digitate*
Bead Tree – *Melia azedarach*
Beef-wood – *Casuarina equisetifolia*
Benjamin Fig – *Ficus benjamina*
Berlin Weed – *Ficus elastica*
Be-still Tree – *Thevetia peruviana*
Bittersweet Orange – *Citrus aurantium × sinensis*
Black Locust – *Robinia pseudacacia*

Black Mulberry – *Morus nigra*
Black Poplar – *Populus nigra*
Black Wattle – *Acacia decurrens*
Blue Gum – *Eucalyptus globulus*
Blue-leaved Wattle – *Acacia cyanophylla*
Blue Mahoe – *Hibiscus elatus*
Bo Tree – *Ficus religiosa*
Botany Bay Fig – *Ficus rubiginosa*
Bottle Tree – *Brachychiton diversifolium*
Bow Wood – *Maclura pomifera*
Box Elder (Maple) – *Negundo aceroides*
Brazilian Pepper Tree – *Schinus terebinthifolius*
Bread-fruit Tree – *Artocarpus altilis*
Brezo – *Erica arbora*
Bull Bay – *Magnolia grandiflora*
Bush Yate – *Eucalyptus lehmannii*
Bushy Blue Gum – *Eucalyptus* × sp.
Butter Bush – *Pittosporum phillyraeoides*
Butternut – *Juglans cinerea*
Calabash Nutmeg – *Monodora myristica*
Calabash Tree – *Crescentia cujete*
California Pepper tree – *Schinus molle*
Campbell's Magnolia – *Magnolia campbellii*
Camel-thorn – *Acacia giraffae*
Camphor Tree – *Cinnamomum camphora*
Candleberry Tree – *Aleurites moluccana*
Candlenut Tree – *Aleurites moluccana*
Canistel – *Pouteria campechiana*
Carob Tree – *Ceratonia siliqua*
Cashew Nut – *Anacardium occidentale*
Catalpa – *Catalpa bignonioides*
Cattley Guava – *Psidium cattleianum*
Ceylon Gooseberry – *Dovyalis hebecarpa*
Chaste Tree – *Vitex agnus-castus*
Cherimoya – *Annona cherimolia*
Cherries – *Prunus* spp.
Cherry Laurel – *Laurocerasus officinalis*
Chicle Tree – *Manilkara achras*
Chinaberry Tree – *Melia azedarach*
China Tree – *Melia azedarach*
Chinese Banyan – *Ficus* spp.
Chinese Persimmon – *Diospyros kaki*
Chinese Pistachio – *Pistacia chinensis*
Chinese Weeping Willow – *Salix babylonica*
Chironja – *Citrus paradisi × sinensis*
Christmas Berry Tree – *Schinus terebinthifolius*

Cigar Box Tree – *Cedrela odorata*
Cinnamon Tree – *Cinnamomum zeylanicum*
Citron – *Citrus medica*
Clove Tree – *Syzygium aromaticum*
Clown Fig – *Ficus aspera*
Cock's Comb Tree – *Erythrina crista-galli*
Cocoa Tree – *Theobroma cacao*
Common Fig – *Ficus carica*
Common Guava – *Psidium guajava*
Common Holly – *Ilex aquifolium*
Common Ironwood – *Casuarina equisetifolia*
Common Lilac – *Syringa vulgaris*
Common Locust – *Robinia pseudacacia*
Common Oak – *Quercus robur*
Common Walnut – *Juglans regia*
Cootamundra Wattle – *Acacia baileyana*
Coral Tree – *Erythrina caffra*
Cork Oak – *Quercus suber*
Council Tree – *Ficus altissima*
Cow-tree – *Brosinum galactodendron*
Crack Willow – *Salix fragilis*
Crape Myrtle – *Lagerstroemia speciosa*
Cuba Bark, Cuba Bast – *Hibiscus elatus*
Custard Apple – *Annona reticulata*
Damson Plum – *Prunus insititia*
Date Plum – *Diospyros kaki*
East Indian Walnut – *Albizia lebbek*
Egg-fruit Tree – *Pouteria campechiana*
Elephant Hedge Bean Tree – *Schotia latifolia*
Elephant's Ear – *Enterolobium cyclocarpum*
Elm Tree – *Ulmus minor*
English Holly – *Ilex aquifolium*
English Oak – *Quercus robur*
English Walnut – *Juglans regia*
European Hackberry – *Celtis australis*
European Holly – *Ilex aquifolium*
Exile Oil Tree – *Thevetia peruviana*
False Acacia – *Robinia pseudacacia*
False Olive – *Elaeodendron orientale*
Fever Tree – *Eucalyptus globulus*
Fiddle Fig – *Ficus lyrata*
Fiddle-leaf Fig – *Ficus lyrata*
Fiddlewood – *Citharexylum spinosum*
Fiery Torch – *Spathodea campanulata*
Fire-wheel Tree – *Stenocarpus sinuatus*
Flamboyant – *Delonix regia*

Flame Bottle-tree – *Brachychiton acerifolium*
Flame Tree – *Brachychiton, Delonix*
Florida Strangling Fig – *Ficus aurea*
Floss-silk Tree – *Chorisia speciosa*
Forest Dog Rose – *Dombeya tiliacea*
Fountain Tree – *Spathodea campanulata*
Frangipani – *Plumeria rubra*
Garden Plum – *Prunus domestica*
Genip – *Melicoccus bijugatus*
Glossy Privet – *Ligustrum lucidum*
Golden Dewdrop – *Duranta repens*
Golden Fig – *Ficus aurea*
Golden Rain-tree – *Koelreuteria paniculata*
Gooseberry Guava – *Psidium cattleianum*
Governor's Plum – *Flacourtia indica*
Grapefruit – *Citrus paradisi*
Green Wattle – *Acacia decurrens*
Greenwattle Acacia – *Acacia decurrens*
Grey Poplar – *Populus canescens*
Guanábana – *Annona muricata*
Guaymochil – *Pithecellobium dulce*
Guava – *Psidium guajava*
Gutta-percha Tree – *Palaquium gutta*
High Rubber Tree – *Ficus altissima*
Hog Plum – *Spondias* spp.
Holly Oak – *Quercus ilex*
Honeyberry – *Melicoccus bijugatus*
Honey Locust – *Gleditsia triacanthos*
Hoop Tree – *Melia azedarach*
Horned Gum – *Eucalyptus cornuta*
Horse-bean – *Parkinsonia aculeata*
Horsetail Casuarina – *Casuarina equisetifolia*
Illawarra Fig – *Ficus rubiginosa*
Illawarra Flame – *Brachychiton acerifolium*
Immortelle – *Erythrina variegata*
India Rubber Fig (tree) – *Ficus elastica*
Indian Almond – *Terminalia catappa*
Indian Banyan – *Ficus benghalensis*
Indian Bean – *Catalpa bignonioides*
Indian Coral-bean – *Erythrina variegata*
Indian Coral-tree – *Erythrina variegata*
Indian Laurel – *Ficus microcarpa*
Indian Lilac – *Melia azedarach*
Indian Walnut – *Aleurites moluccana*
Jacaranda – *Jacaranda mimosifolia*
Jak-fruit Tree – *Artocarpus heterophyllus*

Jambolan – *Syzygium cuminii*
Japanese Chestnut – *Castanea crenata*
Japanese Date Plum – *Diospyros kaki*
Japanese Medlar – *Eriobotrya japonica*
Japanese Persimmon – *Diospyros kaki*
Japanese Pittosporum – *Pittosporum tobira*
Japanese Plum – *Eriobotrya japonica*
Java Plum – *Syzygium cuminii*
Java Willow – *Ficus benjamina*
Jerusalem Thorn – *Parkinsonia aculeata*
Kaffierboom – *Erythrina caffra*
Kakee, Kaki – *Diospyros kaki*
Karaka – *Corynocarpus laevigata*
Karoo Thorn – *Acacia karroo*
Kassod Tree – *Cassia siamea*
Kei Apple – *Dovyalis caffra*
Ketambilla – *Dovyalis hebecarpa*
Krishna Fig – *Ficus krishnae*
Kurrajong – *Brachychiton diversifolium*
Large-leaved Lime – *Tilia platyphyllos*
Laurel Bay – *Laurus nobilis*
Laurel Fig – *Ficus microcarpa*
Laurel-leaved Moonseed – *Cocculus laurifolius*
Lemon – *Citrus limon*
Lemonade Tree – *Rhus typhina*
Lemon Verbena – *Lippia triphylla*
Lignum vitae – *Guaiacum officinale*
Lilac – *Melia azedarach*
Lime – *Citrus aurantifolia*
Limelo – *Citrus aurantifolia* × *paradisi*
Litchi – *Litchi chinensis*
Lobster Plant – *Erythrina crista-galli*
Locust Tree – *Ceratonia siliqua*
Lofty Fig – *Ficus altissima*
Lombardy Poplar – *Populus nigra* 'Italica'
London Plane – *Platanus* × *hybrida*
Loquat-leaved Fig – *Ficus afzelii*
Loquat Tree – *Eriobotrya japonica*
Lucky-bean Tree – *Erythrina caffra*
Lyrate-leaved Fig – *Ficus lyrata*
Madras Thorn – *Pithecellobium dulce*
Madroño – *Arbutus unedo*
Mahoe – *Hibiscus elatus*
Malayan Banyan – *Ficus microcarpa*
Mamey – *Mammea americana*
Mammee-apple – *Mammea americana*

Mandarin – *Citrus reticulata*
Mandarin-lime – *Citrus aurantifolia* × *reticulata*
Mangle Bobo – *Bontia daphnoides*
Mango – *Mangifera indica*
Manila Tamarind – *Pithecellobium dulce*
Margosa – *Azadirachta indica*
Mesquite – *Prosopis juliflora*
Mile Tree – *Casuarina equisetifolia*
Milk-tree – *Brosinum galactodendron*
Milo – *Thespesia populnea*
Mock Orange – *Pittosporum undulatum*
Molle – *Schinus molle*
Monkey Pod – *Pithecellobium dulce*
Monkey's Dinner Bell – *Hura crepitans*
Monkey-thorn – *Acacia galpinii*
Moreton Bay Fig – *Ficus macrophylla*
Mosaic Fig – *Ficus aspera*
Mountain Mahoe – *Hibiscus elatus*
Murray Red Gum – *Eucalyptus camaldulensis*
Narrow-leaved Ash – *Fraxinus angustifolia*
Native Daphe – *Pittosporum undulatum*
Neem Tree – *Azadirachta indica*
Netteltree – *Celtis australis*
New Zealand Christmas Tree – *Metrosideros excelsa*
New Zealand Laurel – *Corynocarpus laevigata*
Nondi Flame – *Spathodea campanulata*
Old Calabar Fig – *Ficus afzelii*
Olive Tree – *Olea europaea*
Ombú – *Phytolacca dioica*
Opiuma – *Pithecellobium dulce*
Opoponax – *Acacia farnesiana*
Orchid Tree – *Bauhinia forficata*
Osage Orange – *Maclura pomifera*
Paloverde – *Parkinsonia aculeata*
Paradise Tree – *Melia azedarach*
Paraguay Tea – *Ilex paraguariensis*
Para Rubber Tree – *Hevea brasiliensis*
Peach Tree – *Amygdalus persica*
Peacock Flower – *Delonix regia*
Pear – *Pyrus communis*
Pecan – *Carya illinoensis*
Peepul Tree – *Ficus religiosa*
Pepper Tree – *Schinus molle*
Persian Lilac – *Melia azedarach*
Persian Walnut – *Juglans regia*
Peruvian Mastic – *Schinus molle*

Peruvian Pepper Tree – *Schinus molle*
Pink Ball – *Dombeya × cayeuxii*
Pink Shower – *Cassia grandis*
Pistachio – *Pistacia* spp.
Plane Maple Tree – *Acer pseudoplatanus*
Plums – *Prunus* spp.
Pohutukawa – *Metrosideros excelsa*
Pokeberry – *Phytolacca dioica*
Pomelo – *Citrus grandis*
Portia Tree – *Thespesia populnea*
Port Jackson Fig – *Ficus rubiginosa*
Portugal Laurel – *Laurocerasus lusitanica*
Portuguese Laurel Cherry – *Laurocerasus lusitanica*
Pride of Bolivia – *Tipuana tipu*
Pride of China – *Melia, Koelreuteria*
Pride of India – *Melia, Koelreuteria, Lagerstroemia*
Purple Guava – *Psidium cattleianum*
Queen Crape Myrtle – *Lagerstroemia speciosa*
Queen's Flower Tree – *Lagerstroemia speciosa*
Queensland Nut – *Macadamia integrifolia*
Rain Tree – *Samanea saman*
Rangoon Creeper – *Quisqualis*
Red Gum – *Eucalyptus camaldulensis*
Rose Apple – *Syzygium jambos*
Rose of India – *Lagerstroemia speciosa*
Rosewood Tree – *Jacaranda mimosifolia*
Rough-skinned Lemon – *Citrus limon × medica*
Royal Poinciana – *Delonix regia*
Rubber Fig – *Ficus elastica*
Rusty Fig – *Ficus rubiginosa*
Sandbox Tree – *Hura crepitans*
Sausage Tree – *Kigelia africana*
Scarlet-flowering Gum – *Eucalyptus ficifolia*
Sea-grape – *Coccoloba uvifera*
Seaside Mahoe – *Thespesia populnea*
Seville Orange – *Citrus aurantium*
Seychelles Rosewood – *Thespesia populnea*
Shaddock – *Citrus grandis*
Shaving-brush Tree – *Pachira insignis*
Shea-butter Tree – *Butyrospermum paradoxum*
She-Oak – *Casuarina equisetifolia*
Shingle Oak – *Casuarina stricta*
Silk-cotton Tree – *Ceiba pentandra*
Silk Oak – *Grevillea robusta*
Silver Lime – *Tilia tomentosa*
Silver Oak – *Grevillea robusta*

Silver Wattle – *Acacia dealbata*
Smoke Tree – *Cotinus coggygria*
Smooth Japanese Maple – *Acer palmatum*
Snake Tree – *Ficus elastica*
Soap-berry Tree – *Sapindus saponaria*
Sour Cherry – *Prunus cerasus*
Sour Orange – *Citrus aurantium*
Soursop – *Annona muricata*
South African Syringa – *Melia azedarach*
Southern Magnolia – *Magnolia grandiflora*
South Sea Ironwood – *Casuarina equisetifolia*
Spanish Chestnut – *Castanea sativa*
Spanish Lime – *Melicoccus bijugatus*
Spanish Plane – *Platanus × hybrida*
Spanish Plum – *Spondias* spp.
Spotted Fig – *Ficus virens*
Staghorn Tree – *Rhus typhina*
Star Apple – *Chrysophyllum cainito*
St.Johns' Bread – *Ceratonia siliqua*
Strangler Fig – *Ficus aurea*
Strawberry Guava – *Psidium cattleianum*
Strawberry Tree – *Arbutus unedo*
Sugar-apple – *Annona squamosa*
Sumach – *Rhus coriaria*
Susanna – *Citharexylum spinosum*
Swamp Yate – *Eucalyptus occidentalis*
Sweet Acacia – *Acacia farnesiana*
Sweet Bean – *Gleditsia triacanthos*
Sweet Cherry – *Prunus avium*
Sweet Lime – *Citrus aurantifolia × medica*
Sweet Locust – *Gleditsia triacanthos*
Sweet Orange – *Citrus sinensis*
Sweet-thorn – *Acacia karroo*
Sycamore – *Acer pseudoplatanus*
Sydney Golden Wattle – *Acacia longifolia*
Talha – *Acacia raddiana*
Tamarind – *Tamarindus indica*
Tangelo – *Citrus paradisi × reticulata*
Tangerine – *Citrus reticulata*
Tangor – *Citrus reticulata × sinensis*
Tasmanian Blue Gum – *Eucalyptus globulus*
Teak Wood – *Tectona grandis*
Texas Umbrella Tree – *Melia azedarach* var.
Thread-leaved Grevillea – *Grevillea nematophylla*
Three-thorned Acacia – *Gleditsia triacanthos*
Tiger's Claw – *Erythrina variegata*

Tipa, Tipuana – *Tipuana tipu*
Tree Heath – *Erica arborea*
Tree of Heaven – *Ailanthus altissima*
Tree of Life – *Guaiacum officinale*
Tropical Almond – *Terminalia catappa*
Trumpet Tree – *Tabebuia rosea*
Tuerckheim's Fig – *Ficus tuerckheimii*
Tulip Tree – *Liriodendron tulipifera*
Umbrella Thorn – *Acacia raddiana*
Umbrella Tree – *Melia azedarach*
Vada Tree – *Ficus benghalensis*
Varnish Tree – *Aleurites, Koelreuteria*
Victorian Box – *Pittosporum undulatum*
Victorian Laurel – *Pittosporum undulatum*
Weeping Fig – *Ficus benjamina*
Weeping Laurel – *Ficus benjamina*
Weeping Pepper – *Schinus molle*
Weeping Pittosporum – *Pittosporum phillyraeoides*
Weeping Willow – *Salix* spp.
West Indian Cedar – *Cedrela odorata*
West Indian Lilac – *Melia azedarach*
White Laurel Magnolia – *Magnolia grandiflora*
White Mulberry – *Morus alba*
White Poplar – *Populus alba*
White Sapote – *Casimiroa edulis*
White Willow – *Salix alba*
White Wood – *Lagunaria patersonii*
Whistling Pine – *Casuarina equisetifolia*
Wild Breadnut – *Pachira insignis*
Wild Chestnut – *Pachira insignis*
Willow – *Casuarina equisetifolia*
Willow Fig – *Ficus benjamina*
Woman's Tongue – *Albizia lebbek*
Yellow Oleander – *Thevetia peruviana*
Yellow Shower – *Cassia spectabilis*
Yerba Mate – *Ilex paraguariensis*
Ylang-Ylang – *Cananga odorata*

Common Names – II

(Some Spanish or Canarian names)

Acacia – *Acacia, Gleditsia, Robinia*
Acacia azul – *Acacia cyanophylla*
Acacia de tes espinas – *Gleditsia triacanthos*
Acacia negra – *Acacia melanoxylon*
Acebo – *Ilex aquifolium*
Adelfa amarilla – *Thevetia peruviana*
Aguacate – *Persea americana*
Ailanto – *Ailanthus altissima*
Alamo blanco – *Populus alba, canescens*
Albaricoque – *Armeniaca vulgaris*
Albizia – *Albizia* spp.
Alcornoque – *Quercus suber*
Algarrobo – *Ceratonia siliqua*
Aligustre – *Ligustrum lucidum*
Almendro – *Amygdalus communis*
Almendro tropical – *Terminalia catappa*
Alméz – *Celtis australis*
Anón – *Annona reticulata*
Apodia – *Spathodea campanulata*
Arbol bonito – *Ficus macrophylla*
Arbol calabaza – *Crescentia cujete*
Arbol candil – *Aleurites moluccana*
Arbol coral – *Erythrina* spp.
Arbol de alcanfor – *Cinnamomum camphora*
Arbol de caucho – *Ficus elastica*
Arbol de la canela – *Cinnamomum zeylanicum*
Arbol de la llama – *Brachychiton acerifolium*
Arbol de las Pagodas – *Ficus religiosa*
Arbol de la vida – *Guaiacum officinale*
Arbol de los Dioses – *Ailanthus altissima*
Arbol del Paraíso – *Melia azedarach*
Arbol jabón – *Sapindus, Pistacia*
Arbol orquídea – *Chorisia, Bauhinia*
Arbol salchicha – *Kigelia africana*

Arce – *Acer* spp.
Arce negundo – *Negundo aceroides*
Atemoya – *Annona cherimolia × squamosa*
Azahar de la China – *Pittosporum tobira*
Azarero – *Pittosporum undulatum*
Bauhinia blanca – *Bauhinia forficata*
Bellasombra – *Citharexylum, Phytolacca*
Benjamina – *Ficus benjamina*
Bombeya – *Dombeya × cayeuxii*
Braquiquito – *Brachychiton diversifolium*
Brezo – *Erica arborea*
Canelo – *Cinnamomum zeylanicum*
Canistel – *Pouteria campechiana*
Caqui – *Diospyros kaki*
Cassia – *Cassia spectabilis*
Castaño – *Castanea sativa*
Casuarina – *Casuarina* spp.
Catalpa – *Catalpa bignonioides*
Cattleya – *Psidium cattleianum*
Cedro – *Cedrela odorata*
Ceibo – *Erythrina crista-galli*
Cerezo – *Prunus avium*
Cerezo laurel – *Laurocerasus lusitanica*
Cherimoya – *Annona cherimolia*
Chopo – *Populus nigra, Ulmus minor*
Cidro – *Citrus medica*
Cina-cina – *Parkinsonia aculeata*
Ciruelo – *Prunus domestica*
Coco nuéz – *Aleurites moluccana*
Cornudo – *Eucalyptus cornuta*
Cresta-gallo – *Erythrina crista-galli*
Dombeya – *Dombeya × cayeuxii*
Duranta – *Duranta repens*
Durazno – *Amygdalus persica*
Encina – *Quercus ilex*
Eritrina – *Erythrina* spp.
Especiero – *Schinus molle*
Eucalipto – *Eucalyptus* spp.
Falsa acacia – *Robinia pseudacacia*
Falsa naranja – *Maclura pomifera*
Falso laurel – *Cocculus laurifolius*
Falso olivo – *Elaeodendron orientale*
Falso plátano – *Acer pseudoplatanus*
Ferruginoso – *Ficus rubiginosa*
Ficus, Fisco – *Ficus elastica*
Flamboyán – *Delonix regia*

Flor de cera – *Plumeria rubra*
Frangipani – *Plumeria rubra*
Fresno – *Fraxinus angustifolia*
Gabón – *Spathodea campanulata*
Genipa – *Melicoccus bijugatus*
Glóbulo – *Eucalyptus globulus*
Gomero – *Ficus* spp.
Grape-fruit – *Citrus paradisi*
Guanábana – *Annona muricata*
Guayabo – *Psidium guajava*
Guayabo pequeño – *Psidium cattleianum*
Guayacán – *Guaiacum officinale*
Guaymochil – *Pithecellobium dulce*
Guindo – *Prunus cerasus*
Hibisco arbóreo – *Hibiscus elatus*
Higuera (común) – *Ficus carica*
Higuera abigarrada – *Ficus aspera*
Higuera bengál – *Ficus benghalensis*
Higuera de los Templos – *Ficus religiosa*
Jabilla – *Hura crepitans*
Jaboncilla – *Sapindus saponaria*
Jabonero – *Pistacia chinensis*
Jacaranda – *Jacaranda mimosifolia*
Jambolán – *Syzygium cuminii*
Kakí – *Diospyros kaki*
Lagunaria – *Lagunaria patersonii*
Latonero – *Celtis australis*
Laurel – *Laurus nobilis*
Laurel de la India – *Ficus microcarpa*
Laurel de Nueva Zelanda – *Corynocarpus laevigatus*
Laurel de Portugal – *Laurocerasus lusitanica*
Limas – *Citrus aurantifolia*
Limonero – *Citrus limon*
Lluvia de oro – *Koelreuteria paniculata*
Macadamia – *Macadamia integrifolia*
Madroño – *Arbutus unedo*
Magnolia – *Magnolia grandiflora*
Magnolia rosada – *Magnolia campbellii*
Majagüilla – *Thespesia populnea*
Mamey – *Mammea americana*
Mamón – *Melicoccus bijugatus*
Mandarina – *Citrus reticulata*
Mango – *Mangifera indica*
Manzano – *Malus pumila*
Margosa – *Azadirachta indica*
Melocotón – *Amygdalus persica*

Metrosidero – *Metrosideros excelsa*
Milo – *Thespesia populnea*
Mimbrera – *Salix fragilis*
Mimosa – *Acacia, Albizia*
Mimosa llorón – *Acacia saligna*
Mimosa picuda – *Acacia karroo, raddiana*
Mimosa plateada – *Acacia dealbata*
Molle – *Schinus molle*
Momona – *Annona cherimolia*
Moral – *Morus nigra*
Morera – *Morus alba*
Naranjero – *Citrus sinensis*
Naranjo amargo – *Citrus aurantium*
Negundo – *Negundo aceroides*
Nisperero – *Eriobotrya japonica*
Níspero del Japón – *Eriobotrya japonica*
Nogál – *Juglans regia*
Nogal de América – *Carya illinoensis*
Olivo – *Olea europaea*
Olivo bastardo – *Bontia daphnoides*
Olmo – *Ulmus minor*
Ombú – *Phytolacca dioica*
Opiuma – *Pithecellobium dulce*
Pagoda – *Ficus religiosa*
Palo borracho – *Chorisia speciosa*
Pandurata – *Ficus lyrata*
Paraíso – *Melia azedarach*
Parkinsonia – *Parkinsonia aculeata*
Pecán – *Carya illinoensis*
Peral – *Pyrus communis*
Pica-pica – *Lagunaria patersonii*
Pimentero – *Schinus molle*
Pino de oro – *Grevillea robusta*
Pino marítimo – *Casuarina* spp.
Plátano (del Líbanon) – *Platanus × hybrida*
Pomarosa – *Syzygium jambos*
Quetembilla – *Dovyalis hebecarpa*
Quinipa – *Melicoccus bijugatus*
Reina de las flores – *Lagerstroemia speciosa*
Riñón – *Annona squamosa*
Roble – *Quercus robur*
Roble blanco – *Tabebuia rosea*
Sao – *Salix* spp.
Sapote blanco – *Casimiroa edulis*
Sauce – *Salix* spp.
Sauce blanco – *Salix alba*

Sauce llorón – *Salix babylonica* et sp.
Tala blanca – *Duranta repens*
Tamarindo – *Tamarindus indica*
Tangelo – *Citrus paradisi* × *reticulata*
Tangor – *Citrus reticulata* × *sinensis*
Tevetia – *Thevetia peruviana*
Tilo – *Tilia platyphyllos*
Tipa – *Tipuana tipu*
Toronja – *Citrus grandis*
Totumo – *Crescentia cujete*
Tulipero del Gabón – *Spathodea campanulata*
Turbito – *Schinus terebinthifolius*
Uva del mar – *Coccoloba uvifera*
Zapatero – *Tabebuia rosea*
Zapote blanco – *Casimiroa edulis*

List of trees native in the Canary Islands

E = endemic in the Canary Islands; M = Macaronesian endemic;
X = sometimes shrubby; ? = originally introduced

M *Apollonias barbujana* (Cav.) Bornm. – Barbusano negro – Lauraceae
E *Apollonias ceballosi* Sventenius – Barbusano blanco – Lauraceae
E *Arbutus canariensis* Veill. – Madroño – Ericaceae
M *Ardisia bahamensis* (Gaertn.) DC. – Aderno, Sacatero – Myrsinaceae
M *Dracaena draco* (L.) L. – Drago – Agavaceae
X *Erica arborea* L. – Brezo – Ericaceae
Erica scoparia L.
E ssp. *platycodon* (Webb & Berth.) Hansen & Kunkel – Tejo, Flejo – Ericaceae
Ilex canariensis Poir.
M ssp. *canariensis* – Acebiño – Aquifoliaceae
M ssp. *azevinho* (Loes.) Kunkel
Ilex platyphylla Webb & Berth.
E ssp. *platyphylla* – Naranjo salvaje – Aquifoliaceae
E ssp. *lopezlilloi* Kunkel
E *Juniperus cedrus* Webb & Berth. – Cedro – Cupressaceae
Juniperus phoenicea L. – Sabina – Cupressaeae
Laurocerasus lusitanica (L.) Roem.
E ssp. *hixa* (Willd.) Kunkel – Hija – Rosaceae
M *Laurus azorica* (Seub.) Franco – Loro, Laurel – Lauraceae
E *Maytenus canariensis* (Loes.) Kunkel & Sunding – Peralillo – Celastraceae
Myrica faya Aiton – Faya – Myricaceae
M *Ocotea foetens* (Ait.) Benth. & Hook.f. – Til – Lauraceae
Olea europaea L. – Oleaceae
E ssp. *cerasiformis* (Webb & Berth.) Kunkel & Sunding – Acebuche
M *Persea indica* (L.) Spreng. – Viñátigo – Lauraceae
E *Phoenix canariensis* hort. ex Chab. – Palmera canaria – Arecaceae
? *Phoenix dactylifera* L. – Datilera – Arecaceae
M *Picconia excelsa* (Ait.) DC. – Palo blanco – Oleaceae
E *Pinus canariensis* Chr.Sm. ex DC. – Pino canario – Pinaceae
Pistacia atlantica Desf. – Almácigo – Pistaciaceae
X *Pistacia lentiscus* L. – Lentisco – Pistaciaceae
E *Pleiomeris canariensis* (Willd.) A.DC. – Coderno, Marmulán – Myrsinaceae

M *Rhamnus glandulosa* Aiton – Sanguino – Rhamnaceae
Salix canariensis Chr.Sm. ex Link – Sao – Salicaceae
M *Sideroxylon marmulano* Banks ex Lowe – Marmulano – Sapotaceae
X *Tamarix africana* Poir. – Tarajal (negro) – Tamaricaceae
X *Tamarix canariensis* Willd. – Tarajal – Tamaricaceae
M *Visnea mocanera* L.f. – Mocán – Theaceae

Other large shrubs, often small trees, include species of

Adenocarpus (Fabaceae)
Anagyris (Fabaceae)
Bencomia (Rosaceae)
Bosea (Amaranthaceae)
Chamaecytisus (Fabaceae)
Convolvulus (Convolvulaceae)
Echium (Boraginaceae)
Euphorbia (Euphorbiaceae)
Gesnouinia (Urticaceae)
Marcetella (Rosaceae)
Phillyrea (Oleaceae)
Sambucus (Sambucaceae)
Spartocytisus (Fabaceae)
Viburnum (Caprifoliaceae)
Withania (Solanaceae)
and others.

REFERENCES

Adams, C. D. (et al.), 1972: Flowering Plants of Jamaica. – University of the West Indies/Glasgow University Press.

Airy Shaw, H. K., 1973 (see J. C. Willis).

Alberti, F. R., 1959: Aceráceas. – Vol. VII, fasc. 115 of Las Plantas Cultivadas en la República Argentina; Buenos Aires.

Aubreville, A. (2nd ed.) 1959: La Flore Forestière de la Côte d'Ivoire. Vols. 1–3. – Centre Technique Forestier Tropical; Nogent-sur-Marne.

Ball, P. W., 1968: Eriobotrya (p. 71), Ceratonia (p. 83), Gleditsia (p. 84), Robinia (p. 106) in Tutin et al. (eds.): Flora Europaea, vol. 2; Cambridge University Press, London.

Barrett, M. F., 1947: Ficus in Florida. I. Australian Species. – American Midland Naturalist 36: 412–430.

Barrett, M. F., 1948: Ficus in Florida. II. African Species. – American Midland Naturalist 39: 188–219.

Barrett, M. F., 1951: Ficus in Florida. III. Asiatic Species. – American Midland Naturalist 45: 118–183.

Barrett, M. F., 1956: Common Exotic Trees of South Florida (Dicotyledons). – University of Florida Press, Gainsville.

Bean, W. J. (rev.ed. by Sir George Taylor et al.), 1970, 1973; Trees and Shrubs Hardy in the British Isles. – John Murray, London.

Beyron, U., 1969: Kyskhetsträd och Änglatrumpeter. – Natur och Kultur, Stockholm.

Bircher, W. H., 1960: Gardens of the Hesperides. – The Anglo-Egyptian Bookshop, Cairo.

Blackwell, W. H., 1968: Flora of Panama. Part VIII. Sapotaceae. – Annals Missouri Botanical Garden 55: 145–169.

Blackwell, W. H. & C. H. Dodson, 1968: Flora of Panama. Part VI. Anacardiaceae. – Annals Missouri Botanical Garden 54: 351–379.

Browicz, K., 1968: Tilia. pp. 247–248 in Tutin et al. (eds): Flora Europaea, vol. 2; Cambridge University Press, London.

Burger, W., 1971: Flora Costaricensis. Family 40. Casuarinaceae. – Fieldiana, Bot. 35: 3–4.

Burges, N. A., 1968: Eucalyptus. pp. 304–305 in Tutin et al. (eds.): Flora Europaea, vol. 2; Cambridge University Press, London.

Burkill, I. H. (et al.), 1966: A Dictionary of the Economic Products of the Malay Peninsula. 2 vols. – Ministry of Agriculture and Co-operation, Kuala Lumpur.
Calabria, J. M., 1968: Frutales de Venezuela. – Fundación Eugenio Mendoza, Caracas.
Carbo, A. & O. Vidal, 1976: El Caqui. – Hojas Divulgadoras 7–76; Ministerio de Agricultura, Madrid.
Carpenter, J. B. & P. C. Reece, 1969: Catalog of Genera, Species and Subordinate Taxa in the Orange Subfamily Aurantioideae (Rutaceae). – USDA, Crops Research Division, ARS 34–106; Beltsville.
Chanes, R., 1969: Deodendron. Arboles y Arbustos de Jardín en Clima Templado. – Editorial Blume, Barcelona.
Chittenden, F. J. (ed.; rev.ed., 4 vols., by P. M. Synge et al.), 1965: Dictionary of Gardening. – Royal Horticultural Society/Clarendon Press.
Cobley, L. S. (rev.ed.) 1965: An Introduction to the Botany of Tropical Crops. – Longmans, Green & Co., London.
Condit, I. J., 1955: Fig varieties, a Monograph. – Hilgardia 23: 323–538.
Condit, I. J., 1969: Ficus, the Exotic Species. – University of California, Division of Agricultural Sciences.
Corner, E. J. H. (2nd ed.) 1952: Wayside Trees of Malaya. 2 vols. – Government Printing Office, Singapore.
Corner, E. J. H. & W. T. Stearn, 1969 (see P. M. Synge).
Degener, O., 1949: Tropical Plants the World Around. – Journal New York Botanical Garden, Apr.: 74–91, May: 110–125.
Dimitri, M. J. & F. R. Alberti, 1952: Tiliáceas. – Vol. VII, fasc. 123 of Las Plantas Cultivadas en la República Argentina; Buenos Aires.
Dimitri, M. J. & V. A. Milano, 1950: Fagáceas. – Vol. IV, fasc. 54 of Las Plantas Cultivadas en la República Argentina; Buenos Aires.
Dimitri, M. J. & V. A. Milano, 1951: Juglandáceas. – Vol. IV, fasc. 52 of Las Plantas Cultivadas en la República Argentina; Buenos Aires.
Ehrendorfer, F., 1971: Spermatophyta, in Strasburger: Lehrbuch der Botanik, 30th ed.; Gustav Fischer Verlag, Stuttgart.
El Hadidi, M. N. & L. Boulos, 1968: Street Trees in Egypt. – Publication Cairo Herbarium N° 1.
Eliovson, S. (6th ed.) 1969: Flowering Shrubs, Trees and Climbers for Southern Africa. – Howard Timmins, Cape Town.
Elwes, H. J. & A. H. Henry (rep.ed.) 1970: The Trees of Great Britain and Ireland. – S. R. Publs. Ltd./Royal Forestry Society.
Encke, F. (ed.; rev.ed.) 1958–1961: Pareys Blumengärtnerei. 3 vols. – Verlag Paul Parey, Berlin & Hamburg.
Esdorn, I. & H. Pirson (2nd ed.) 1973: Die Nutzpflanzen der Tropen und Subtropen in der Weltwirtschaft. – Gustav Fischer Verlag, Stuttgart.
Fabris, H. A., 1969: Bigoniáceas. – Vol. X, fasc. 173 of Las Plantas Cultivadas en la República Argentina; Buenos Aires.
Fitschen, J. (5th ed. by F. Boerner), 1959: Gehölzflora. – Quelle & Meyer, Heidelberg.

Flinta, C. M., 1970: Peru and Bolivia. pp. 347–366 in R. N. Kaul (ed.): Afforestation in Arid Zones; Dr. W. Junk, Publ., The Hague.
Font Quer, P. (2nd ed.) 1973: Plantas Medicinales. El Dioscórides Renovado. – Editorial Labor, Barcelona.
Fosberg, F. R., Falanruw, M. V. C. & M.-H. Sachet, 1975: Vascular Flora of the Northern Marianas Islands. – Smithsonian Contributions to Botany N° 22.
Fosberg, F. R. & M.-H. Sachet, 1972: Thespesia populnea (L.) Solander ex Correa and Thespesia populneoides (Roxburgh) Kosteletsky (Malvaceae). – Smithsonian Contributions to Botany N° 7.
Franco, J. do A., 1964: Populus (pp. 54–55), Pittosporum (pp. 383–384) in Tutin et al. (eds.): Flora Europaea, vol. 1; Cambridge University Press, London.
Franco, J. do A., 1968: Acacia (pp. 84–85) in Tutin et al. (eds.): Flora Europaea, vol. 2; Cambridge University Press, London.
Franco, J. do A. & M. L. da Rocha Afonso, 1972: Fraxinus, Olea (pp. 53–55) in Tutin et al. (eds.): Flora Europaea, vol. 3; Cambridge University Press, London.
Garcia C., A., 1961: Botanical Gardens of Orotava. Descriptive Guide. – Orotava, Tenerife.
Gentry, A. H., 1973: Flora of Panama. Part IX. Bignoniaceae. – Annals Missouri Botanical Garden 60: 781–977.
Goor, A. Y. & C. W. Barney, 1968: Forest Tree Planting in Arid Zones. – Ronald Press Co., New York.
Graf, A. B. (rev.ed.) 1963: Exotica 3. Pictorial Cyclopedia of Exotic Plants. – Roehrs Co., Publs., Rutherford.
Grant, M. L., Fosberg, F. R. & H. M. Smith, 1974: Partial Flora of the Society Islands: Ericaceae to Apocynaceae. – Smithsonian Contributions to Botany N° 17.
Hamlin, B., 1962: Nature in New Zealand: Native Trees. – A. H. & A. W. Reed, Publs., Wellington.
Harrison, R. E. (rev.ed.) 1960: Handbook of Trees and Shrubs for the Southern Hemisphere. – Harrison & Reed, Palmerston North & Wellington.
Hart, C. & C. Raymond (2nd ed.) 1974: British Trees in Colour. – Michael Joseph, London.
Hegi, G. (et al.), 1906–1931: Illustrierte Flora von Mitteleuropa. – Carl Hanser Verlag, München.
Hutchinson, J., 1967: Key to the Families of Flowering Plants of the World. – Clarendon Press, Oxford.
Huxley, A., 1974: Plant and Planet. – Allen Lane, London.
Irvine, F. R. (et al.), 1961: Woody Plants of Ghana, with Special Reference to their Use. – Oxford University Press, London.
Jaynes, R. A. (ed.), 1969: Handbook of North American Nut Trees. – Northern Nut Growers Association, Knoxville.

Kaul, R. N. (ed.), 1970: Afforestation in Arid Zones. – Monographiae Biologicae 20; Dr. W. Junk, Publs., The Hague.
Keay, R. W. J., Onochie, C. F. A. & D. P. Stanfield, 1964: Nigerian Trees. 2 vols. – Department of Forest Research, Ibadan.
Kennard, W. C. & H. F. Winters, 1960: Some Fruits and Nuts for the Tropics. – USDA, Miscell. Publication N° 801, Washington.
Kester, D. E., 1969: Almonds. pp. 302–314 in R. A. Jaynes (ed.): Handbook of North American Nut Trees; Northern Nut Growers Association, Knoxville.
Kunkel, G., 1969: Arboles Exóticos. Los Arboles Cultivados en Gran Canarias. – Edic. Excmo. Cabildo Insular de Gran Canaria, Las Palmas.
Kunkel, G., 1970: Las Plantas Alimenticias y Especieras de las Islas Canarias. – Jardineria en Canarias N° 1; Las Palmas.
Kunkel, M. A. & G. Kunkel, 1974: Flora de Gran Canaria. Vol. 1: Arboles y Arbustos Arbóreos. – Ediciones Excmo. Cabildo Insular de Gran Canaria, Las Palmas.
Ledin, R. B. & E. A. Menninger, 1956: Bauhinia, the so-called Orchid Trees. – National Horticultural Magazine 35: 183–200.
Little, E. L. & F. H. Wadsworth, 1964: Common Trees of Puerto Rico and the Virgin Islands. – USDA, Agricultural Handbook N° 249; Washington.
Little, E. L., Woodbury, R. O. & F. H. Wadsworth, 1974: Common Trees of Puerto Rico and the Virgin Islands. – USDA, Agricultural Handbook N° 449, Washington.
Lombardo, A., 1958: Los Arboles Cultivados en los Paseos Públicos. – Concejo Departamental, Montevideo.
Lombardo, A., 1961: Los Arbustos y Arbustillos de los Paseos Públicos. – Concejo Departamental, Montevideo.
Lombardo, A. (2nd ed.) 1964: Flora Arbórea y Arborescente del Uruguay Concejo Departamental, Montevideo.
Long, R. W. & O. Lakela, 1971: A Flora of Tropical Florida. – University of Miami Press, Coral Gables.
López P., J., 1968: Variedades de Almendras. – pp. 103–125 in S. Alvarez et al.: Diez Temas sobre Frutos Secos; Publicación de Capacitación Agraria, Madrid.
Madden, G., Brison, F. R. & J. M. McDaniel, 1969: Pecans. – pp. 163–189 in R. A. Jaynes (ed.): Handbook of North American Nut Trees; Northern Nut Growers Association, Knoxville.
Malo, S. E., 1975: Manual del Cultivo del Aguacate. – Servicio Agrícola, Caja Insular de Ahorros, Las Palmas.
Marthi, C. E., 1958: Pitosporáceas. – Vol. V., fasc. 90 of Las Plantas Cultivadas en la República Argentina; Buenos Aires.
Martinez, M., 1969: Las Plantas Medicinales de Mexico. – Ediciones Botas, Mexico.
Marzocca, A., 1950: Ebenáceas. – Vol. VIII, fasc. 158 of Las Plantas Cultivadas en la República Argentina; Buenos Aires.

Marzocca, A., 1952: Apocináceas. – Vol. IX, fasc. 163 of Las Plantas Cultivadas en la República Argentina; Buenos Aires.
McClintock, E., 1953: The Cultivated Species of Erythrina. – Baileya 1: 53–58.
McKay, J. W. & R. A. Jaynes, 1969: Chestnuts. – pp. 264–286 in R. A. Jaynes (ed.): Handbook of North American Nut Trees; Northern Nut Growers Association, Knoxville.
Menninger, E. A., 1949: Evergreen Trees for Street Planting in Warm Regions. – Proceedings Florida State Horticultural Society LXII: 189–196.
Menninger, E. A., 1959: The Cultivated Eugenias in American Gardens. – National Horticultural Magazine 38: 92–104, 145–163.
Menninger, E. A., 1960: Tabebuia, our best Yard Trees. – Proceedings Florida State Horticultural Society 73: 366–373.
Menninger, E. A., 1962: Flowering Trees of the World for Tropics and Warm Climates. – Heartside Press, Inc.; New York.
Menninger, E. A., 1964: Seaside Plants of the World. – Hearthside Press, Inc.; New York.
Menninger, E. A., 1967: Fantastic Trees. – Viking Press, New York.
Milano, V. A., 1964: Fitolacáceas. – Vol. IV, fasc. 66 of Las Plantas Cultivadas en la República Argentina; Buenos Aires.
Milano, V. A. & E. P. Molinari, 1963: Plantanáceas. – Vol. V, fasc. 92 of Las Plantas Cultivadas en la República Argentina; Buenos Aires.
Mitchell, A., 1974: A Field Guide to the Trees of Britain and Northern Europe. – Collins, London.
Moeller, H., 1968: What's Blooming Where on Tenerife?. – Bambi-Verlag, Pto. de la Cruz, Tenerife.
Moeller, H., 1971: Kanarische Pflanzenwelt II. – Bambi-Verlag, Pto. de la Cruz, Tenerife.
Moggi, G., 1964: Guida al Riconoscimento degli Eucalitti coltivati in Italia. – Pubbl. Centro Sper. Agric. For. 7: 147–220.
Moldenke, H. N., 1973: Flora of Panama. Part IX. Verbenaceae. – Annals Missouri Botanical Garden 60: 41–148.
Molesworth Allen, B., 1967: Malayan Fruits. An Introduction to the Cultivated Species. – Donald Moore Press, Ltd.; Singapore.
Molinari, E. P., 1965: Malváceas. – Vol. VII, fasc. 124 of Las Plantas Cultivadas en la República Argentina; Buenos Aires.
Molinari, E. P. & V. A. Milano, 1958: Esterculiáceas. – Vol. VII, fasc. 126 of Las Plantas Cultivadas en la República Argentina; Buenos Aires.
Mortensen, E. & E. T. Bullard (rev.ed.) 1968: Handbook of Tropical and Subtropical Horticulture. – USDS, Agency for International Development, Washington.
Neal, M. C. (rev.ed.) 1965: In Gardens of Hawaii. – B. P. Bishop Museum, Special Publication N° 50, Honolulu.
Nowicke, J. W., 1970: Flora of Panama. Part VIII. Apocynaceae. – Annals Missouri Botanical Garden 57: 59–130.

O'Rourke, F. L. S., 1969: The Carpathian (Persian) Walnut. – pp. 232–239 in R. A. Jaynes (ed.): Handbook of North American Nut Trees; Northern Nut Growers Association, Knoxville.
Ozenda, P., 1958: Flore du Sahara Septentrional et Central. – Centre National Recherche Scientifique, Paris.
Palmer, E., Pitman, N. et al., 1972: Trees of Southern Africa. – Vols. I–III, A. A. Balkema, Cape Town.
Parham, B. E. V. & A. J. Healy, 1976: Common Weeds in New Zealand. – New Zealand DSIR Information Series N° 112, Wellington.
Pennington, T. D. & B. T. Styles, 1975: A Generic Monograph of the Meliaceae. – Blumea 22: 419–540.
Perry, F. et al., 1972: Flowers of the World. – Hamlyn, London-New York-Sydney-Toronto.
Pesman, M. W., 1962: Meet Flora Mexicana. – Dale S. King, Publs., Globe, Arizona.
Polunin, O. (2nd ed.) 1974: The Concise Flowers of Europe. – Oxford University Press, London.
Polunin, O. & A. Huxley, 1965: Flowers of the Mediterranean. – Chatto & Windus, London.
Polunin, O. & B. E. Smythies, 1973: Flowers of South-west Europe. A Field Guide. – Oxford University Press, London.
Pope, W. T., 1968: Manual of Wayside Plants of Hawaii. – Charles E. Tuttle Co., Publs., Rutland & Tokyo.
Pryor, L. D., 1976: The Biology of the Eucalypts. – Studies in Biology N° 61; Edward Arnold (Publs.) Ltd., London.
Pryor, L. D. & L. A. S. Johnson, 1971: A Classication of the Eucalypts. – Australian National University, Canberra.
Purseglove, J. W., 1968: Tropical Crops. Dicotyledons 1–2. – Longmans, Green & Co., London & Harlow.
Rauh, W. (2nd ed.) 1950: Morphologie der Nutzpflanzen. – Quelle & Meyer, Heidelberg.
Rechinger, K.-H., 1964: Salix. pp. 43–54 in Tutin et al. (eds.): Flora Europaea, vol. 1; Cambridge University Press, London.
Rendle, A. B., 1956: The Classification of Flowering Plants. Vols. 1–2. – Cambridge University Press, London.
Ruiz de la Torre, J., 1971: Arboles y Arbustos de la España Peninsular. – Instituto Forestal de Investigaciones y Experiencias, Madrid.
Schaeffer, H.-H. (2nd ed.) 1967: Pflanzen der Kanarischen Inseln/Plants of the Canary Islands. – Kanaren-Verlag, Ratzeburg.
Schmucker, T., 1952: Was ist ein Baum?. – Mitteilungen Deutsche Dendrologische Gesellschaft 57: 14–19.
Schütt, P., 1972: Weltwirtschaftspflanzen. – Verlag Paul Parey, Berlin & Hamburg.
Schwarz, O., 1964: Quercus. pp. 61–64 in Tutin et al. (eds.): Flora Europaea, vol. 1; Cambridge University Press, London.

Serr, E. F., 1969: Persian Walnuts in the Western States. – pp. 240–263 in R. A. Jaynes (ed.): Handbook of North American Nut Trees; Northern Nut Growers Association, Knoxville.
Smith, C. E., 1960: A Revision of Cedrela (Meliaceae). – Fieldiana, Bot. 29: 295–341.
Soukup, J., 1970: Vocabulario de los Nombres Vulgares de la Flora Peruana. – Colegio Salesiano, Lima.
Van der Spuy, U., 1971: South African Shrubs and Trees for the Garden. – Hugh Keartland, Publs., Johannesburg.
Standley, P. S. & L. O. Williams, 1969: Flora of Guatemala 8/4: Apocynaceae. Dogbane Family. – Fieldiana, Bot. 24: 334–407.
Standley, P. S. & L. O. Williams, 1974: Flora of Guatemala 10/3: Bignoniaceae. Bignonia Family. – Fieldiana, Bot. 24: 153–232.
Storey, W. B., 1969: Macadamia. pp. 321–335 in R. A. Jaynes (ed.): Handbook of North American Nut Trees; Northern Nut Growers Association, Knoxville.
Synge, P. M. (ed.) et al. (2nd ed.) 1969: Supplement to the Dictionary of Gardening. – Royal Horticultural Society/Clarendon Press.
Takhtajan, A., 1969: Flowering Plants, Origin and Dispersal. – Oliver & Boyd, Edinburgh.
Tutin, T. G., 1964: Juglands (pp. 56–57), Castanea (p. 61), Celtis, Ulmus (pp. 65–66), Maclura, Morus (p. 66), Ficus (pp. 66–67), Platanus (p. 384) in Tutin et al. (eds.): Flora Europaea, vol. 1; Cambridge University Press, London.
Tutin, T. G., 1968: Simaroubaceae (pp. 230–231), Melia (p. 231), Anacardiaceae (pp. 236–237) in Tutin et al. (eds.): Flora Europaea, vol. 2; Cambridge University Press, London.
Uphof, J. C. T. (rev.ed.) 1968: Dictionary of Economic Plants. – J. Cramer, Lehre.
Walters, S. M., 1968: Acer (pp. 238–239) in Tutin & al. (eds.): Flora Europaea, vol. 2; Cambridge University Press, London.
Webb, D. A., 1968: Prunus (pp. 77–80) in Tutin et al. (eds.): Flora Europaea, vol. 2; Cambridge University Press, London.
Webb, D. A., 1972: Arbutus (p. 11) in Tutin & al. (eds.): Flora Europaea, vol. 3; Cambridge University Press, London.
Webb, D. A. & E. M. Rix, 1972: Erica (pp. 5–8) in Tutin & al. (eds.): Flora Europaea, vol. 3; Cambridge University Press, London.
Webster, G. L. & D. Burch, 1968: Flora of Panama. Part VI. Euphorbiaceae. – Annals Missouri Botanical Garden 54: 211–350.
Wiggins, I. L., Porter, D. M. et al., 1971: Flora of the Galápagos Islands. – Stanford University Press. Stanford.
Willis, J. C. (8th ed. by H. K. Airy Shaw), 1973: A Dictionary of the Flowering Plants and Ferns. – Cambridge University Press, London.

GENERAL INDEX

Family names are given in capital letters; species described in the text are cited in *italics*.